DESIGN CENTERED INTRODUCTION
TO AEROSPACE ENGINEERING

Narayanan M. Komerath

Published by SCV Inc, Johns Creek, Georgia USA 30022
www.stratcepts.com

The author gratefully acknowledges support under the NASA Innovation in Aerospace Instruction Initiative, NASA Grant No. NNX09AF67G. Tony Springer Technical Monitor.

Any opinions, findings, and conclusions or recommendations expressed in this material are those of the author and do not necessarily reflect the views of the National Aeronautics and Space Administration.

Print Version: ISBN 978-1-949335-00-2
Library of Congress Control Number: 2018907580

Dedicated to my Teachers who taught me to dream of what is ultimately needed, and strive to make that possible

Preface

The successful orbiting of the Sputnik spacecraft by the Soviet Union in 1957 probably caused more of a sensation in the United States of America than anywhere else. It was a shock, given the acknowledged superiority of American technology and logistics following World War II and the rapid rise of prosperity. It caused much introspection about the state of science and mathematics in engineering education. 'Rocket Science' came to denote the ultimate in 'brains' and technological excellence. Departments of Aeronautics swiftly changed their names to Aerospace Engineering. State governments poured resources into STEM education and facilities. Enrollment in aerospace engineering curricula soared. These curricula took a sharp turn to focus on engineering science over practice and design.

Nearly 3 decades later, I was sitting in a faculty meeting where the curricular pendulum had swung all the way back. Driven by concerns from U.S industry, the Accreditation Board for Engineering and Technology had imposed quotas for so-called 'design content' versus 'science content'. Administrators rushed to pay obeisance to Capstone Design as the Culmination of the Entire Curriculum. My job on the other hand, was to teach high speed aerodynamics and aerospace propulsion to our large undergraduate classes - both considered to be 'difficult' courses, requiring a fair amount of conceptual understanding as well as mathematical tools. I tried my best to satisfy the 'design content' pressures with plenty of open-ended problems and synthesis versus analysis. This included writing FORTRAN programs to do quite realistic analyses and optimizations. My students would come in looking sleep-deprived, then inform me that "it isn't because of your course, it's this Design course that keeps us up all night". From observing their mode of getting 'organized' for team meetings, this was not very surprising.

Listening to the exhortations at that particular faculty meeting, I had the growing conviction that those who were loudest in praise of Capstone Design were those who had had no occasion in their stellar careers as scientists, to actually consider how airplanes were designed. Their notion was that students would learn the deepest knowledge that we 'subject area' teachers could impart, and then magically apply and synthesize this and out would come a shiny new airplane. As the Launch Window to get in early at the lunch counter queue, was closing rapidly, and my tummy was growling at me, I said in exasperation: "Look, I can teach the first half of Capstone Design as it is taught today, to anyone with a high school education. It's the one course that actually needs no prerequisites". Their looks reminded me of the famous picture in Charles Dickens' Oliver Twist, captioned *The Boy Asks For More*, but it had the desired effect of ending the meeting. I assumed that no one thought about what I had said, and there I was wrong.

A decade later, the aerospace marketplace pendulum had swung back again. Our undergraduate program was less than half it's size of the 1980s. More alarmingly, we were losing our top sophomores to the "enemy" - the other engineering schools. At the end of their second year, they had still not seen any AE courses. Their ME and EE peers in study groups would press them to jump ship to those disciplines. Our administrators were now fully in support of early intervention, and so was born the idea of the early-sophomore Introduction to Aerospace Engineering. The first attempt was by a senior professor of vast industry and government credentials, who went in with his trademark 80 transparencies per class. Color transparencies. The sophomores watched in apparent rapt attention, but reported learning little of tangible value. The next attempt was by a professor of Structures who argued for early Depth. He started the course by deriving the Navier Stokes and other fundamental equations. Over 40 percent of the class promptly dropped out and alumni screamed. I had only lunchtime anecdotes on these exciting events when I was summoned to the only remaining chair in a a tiny conference room for a group meeting with the School Chair already in progress. Memories from Cold War books on Politburo meetings came to mind...

The question from my dear friend was very direct. He was making his case - and had summoned me as a last-ditch attempt to overcome the skepticism that he was getting. He reminded me of that unwise statement of a decade ago. I readily confessed, but stood by my statement. I pointed out that the community was experimenting with various 'active learning' options, but as much as I loved hardware experiments, it would take large resources which the school would not afford, to conduct such a course for our classes. So we had few options left. Ten minutes later, I realized that I had agreed to teach conceptual design to first-year (not sophomore any more!) students. A few hours later the realization sank in, that these were not third-quarter freshmen, but students on their very first day of college in the Fall. No differential equations, no integrals, no derivatives, no statics or dynamics or college chemistry. No engineering drawing.

This book is the product of the course that was developed to meet that horrifying requirement. I did have fabulous support. For my birthday, Padma gave me a small picture book that described Designing The Boeing 777, so I had 'real-world' numbers and pictures. A sympathetic chemistry teacher from Emory University took me aside at a party and gave me detailed advice on why 'First Year Students Are Different' - a truly invaluable Short Course. I used these well. I distilled the content using a few textbooks to convince myself that the job could be done with no resort to Calculus - even the Breguet Range Equation could be taught using a spreadsheet. When I first taught it, several of my senior colleagues populated the back rows of the auditorium, not least in anticipation of the entertainment of watching me fall on my face. My students saved me, as usual. I am sure they believed that they had a crazy instructor when I informed them that inside 6 weeks they were going to be 'flying' their own transonic airliners. But they rose to the challenge brilliantly. We achieved exactly what we planned. It was truly eye-opening to see what our students could do, if we did not waste our time informing them what they could not do.

A couple of years later, the students were assuming that this course had always existed - they had found the website on it (I knew who had developed that, but they did not, nor did they care: in fact some advised me to try and become as kind as the professor who developed that web page, instead of asking them to do all the thinking!). No longer would they be content with designing a transonic airliner like I had learned in Conceptual design as a 1970s undergraduate in Capstone Design. Now they wanted to design a variety of airplanes, including supersonic ones (where I had no secret resources similar to my birthday present that enabled us to 'design' an Extended Range 777!) They also wanted to design space missions - and first year students started joining various Design Competition activities. We had overcome our fears. So can you, dear reader.

Two senior colleagues including the School Chair also adopted the Design-Centered approach, bringing their own experience to bear. Three papers at the annual conferences of the American Society of Engineering Education described our experience. The approach became the center-piece, the portal, to our new Aerospace Digital Library effort to link up the knowledge base, later expanded in the NASA-funded EXTROVERT project to enable our students to soar across disci-plines without fear, and innovate on grand problems. I taught the Design-Centered Introduction for a few years, and then was moved to other courses.

Another 15 years have passed, and a new curriculum is in force, with a new and enthusiastic set of teachers armed with the latest in technology. We now have resources to bring in active learning for all our students, and so the freshman introduction now focuses mostly on such projects as building and testing dual-stage rockets, balsa wood gliders, conducting flight dynamics, structures and materials, and controls experiments, and getting guest lectures on How The Aerospace Program Works, and introductions to various Clubs and Study Abroad Programs. The course has been slashed to just 1 semester credit hour. I was again to teach this: I realized that there was much I needed to learn. At this writing I have survived three semesters of teaching AE1601, and it has been an absolute delight, with all the support that I got from the instructors in the other two sections, and all the Teaching Assistants.

While all this is very exciting and the modern course lived up to its promise to be much fun, I must still think of the instructor (me!) in the upper division aerodynamics and propulsion courses, which are still labeled as 'difficult'. The recent cohorts in those courses appear to come with a more heterogeneous spread of awareness and preparedness for serious engineering courses. So it is time to revive the content of my design-based course, as a background resource to inform the curious student. Perhaps it will help tie all the disciplines of our field together in the pursuit of coherent objectives such as the design of radically new vehicle concepts.

As with any book written by a university professor, I have thousands of students in different parts of the world to acknowledge and thank for their inspiration, efforts and advice. A few deserve special mention. Jason Akovenko, then an undergrad senior pursuing a research Special Problem in my wind tunnel laboratory between Co-Op work quarters with a leading aircraft manufacturer, quiety assured me that I was on the right track, and armed me with a copy of his team's Report from the first quarter course on Aircraft Conceptual Design. So I had a very clear metric to target. My dear then-graduate students Leigh Ann Darden, Catherine Matos, Liliana Villarreal and Richard Ames, alums of our undergraduate program, ordered me to sit down and listen while they explained that my sympathies for the red-eyed freshmen in my 9AM class were misplaced: it was not math homework, but the Fraternity Rush party season that was keeping them awake until 3AM. The young lady who walked into my office asking if she should drop the course because she had had no advanced physics in high school (why? Because she was too busy taking advanced math!) gave me the desperate idea of asking whether she knew what the first and second derivatives of x with respect to time indicated, and how those related to the ideas of velocity and acceleration. Of course she soared to excellence. Her classmate, last seen a few years ago as a high-flying aerospace corporate executive, came into my office to explain that "we like your class but 9AM is awfully early". She returned at the end of the quarter after a brilliant performance to thank me for my drastic advice, and exclaim that going to sleep at 10PM (instead of 3AM) and waking up at 6AM actually worked as a winning habit. And it was either she or another alum who realized on the final exam that she had not copied down a formula to calculate transonic drag rise, but "remembered that it shot up like an exponential function so I will approximate it with this mathematical expression", demonstrating the essence of aerospace innovation through bold, scientific improvisation. Of course Professor Robert Loewy, School Chair for his faith in allowing me to experiment, and then in adopting and adapting my approach to suit his own experience of real industry design practice. Professor Sankar for his unique twist on using this approach. And of course, my dear friend Professor Erian Armanios for summoning me to that meeting in 1997.

There is no pre-requisite for a course based on this book, except the basic expectations of intense thought and effort that come with a good high school education. It has been condensed into a 1.5-day and then a 1-day Short Course/Workshop, presented first to non-aerospace students at a university, and then to new engineers at a major corporation's research center. The last time I taught the course in this form at Georgia Tech, student feedback expressed deep anguish and self-pity at the 'enormously time-consuming amount of calculations needed to do aerospace engineering', at the end of the first 6-week cycle of design. So I asked them to do a second cycle, this time for a hydrogen-powered version of their airliner. They did it in the one week allotted. I asked them the essence of the procedure on the final exam, as one question, for which they had maybe 35 minutes. They did fine. This is the power of learning through iteration, with the human mind fully engaged.

The other reader whom I have in mind is the PhD candidate in engineering or science, who, like my senior colleagues of 30 years ago, has had no occasion to think of how airplanes come together starting with a concept. Do not be shy: outside our disciplines of extreme concentration, most of us are less knowledgeable than high-school students when it comes to being aware of other disciplines. I hope this book removes your fear of flying over the treetops, leaving the jungles of complexity below you.

Welcome to aerospace engineering. Whatever else it may be, it can never be called boring.

Respectfully submitted

Narayanan Komerath
Atlanta, Georgia
August 2018.

Contents

CONTENTS

Chapter 1

Welcome to Aerospace Engineering

1.1 Why Design?

Most pioneers learned the 'disciplines' of engineering through the hard route of designing devices and systems. The cartoons showing the pioneers of our field strapping wings on their backs and running off cliffs are, sadly, not without basis: in those days there was no real alternative in validating a strongly-believed hypothesis or design. Hordes of snickering neighbors, peers, debt-collectors and news reporters no doubt goaded pioneers into such extreme demonstration attempts. We try to capture this spirit, but without the physical risks, by taking the learner through the *process of conceptually designing* a flight vehicle. Most of all, we try to learn that real design of a new concept is never a straight line process: one has to learn from mistakes, a.k.a. *iteration*. The first part of that is to not give up; but instead to learn what one needs to learn, where one can find it. In the process, we explore the gateways to the various disciplines of aerospace engineering, each leading to deep resources of knowledge and experience.

Prepare for liftoff. We assume that you are a first year undergraduate, fresh out of high school. Or a PhD candidate terrified that you may get caught out, lacking 'practical sense' at the Qualifying Examination and likely to blurt out catastrophically wrong values of lift or drag coefficients. Or a professor who has always wanted to figure out how flight vehicles were designed, but too busy and shy to read the huge volumes on Vehicle Design and Multi-Disciplinary Design Optimization (I have not read those either, but my students have..). You will find yourself designing airliners and defining their flight envelopes, as you go through these chapters, follow the exercises and complete a few assignments.

This book is a condensed, refined version of a course that was first developed in the Summer of 1997, and first taught in Fall 1997 at Georgia Tech. In the 1990s, the aerospace industry was in recession after the Cold War ended and the need for large numbers of combat aircraft disappeared. Many students who came to Georgia Tech to study aerospace engineering, found themselves studying with excellent students from other disciplines. There was no introduction to aerospace engineering in the first year of college, because it was assumed that students could not appreciate the issues in aerospace engineering until they had completed several courses in Mathematics, Physics, Chemistry, Statics, Dynamics and Strength of Materials. As they completed the second year, they found themselves having to decide between continuing alone in a school where they had never taken a course, or going with their study-group friends into other disciplines. The School of Aerospace Engineering decided to try introducing the excitement of aerospace engineering in the first year. Several different approaches were tried, but they were either too simplistic (avoiding Math and Physics) or far too complicated. The notion of using aircraft conceptual design as a vehicle to introduce students in their first week of college to aerospace engineering, was rather risky, but our students took to it with amazing enthusiasm. The course was taught several times in the subsequent years. It was condensed into a 1.5-day Workshop for engineering students, and then for new engineers from other disciplines.

1.2   About Your Classmates (and Teachers)

In most engineering schools around the USA, and probably around the world, Introduction to Aerospace Engineering is a subject taught to students in the second or third year (I took it in my 3rd year) of the engineering curriculum. The reason is that most curricula are structured in linear progression, starting with Mathematics, Physics, Chemistry, Engineering Graphics, Computer-based Problem Solving, and moving on to Statics and Dynamics, Strength of Materials, Economics. The trouble with this strategy is that many (most?) of us came to engineering school rather than other choices of college (which all appealed as being 'easier!') because we dream of things that fly. I, for instance, missed out on my opportunities to become a great Cricket Star (my real ambition as I finished high school) because of my propensity to miss catches even while standing 2 feet from the batsman every time I heard a fighter plane roaring overhead and had to look up there instead. So would most of my students, based on what I see of their tastes. One standard 'test' that I administer is to sketch a squiggle on the board, and ask them: "What is that?" Eighty percent of my students respond immediately: "A Space Shuttle Ready For Liftoff" I ask them why - it looks like a squiggle of a truss and a little something next to it. The Space Shuttle is entirely in their brains, not a compliment to my (poor) drawing skills!

So what happens when they put us through the drudgery of these 'basic' courses for two miserable years? These days students have many career choices - and the ones with the most choices often drop out and go elsewhere! This is what drove us at Georgia Tech, and some others like us, to 'bite the bullet' and develop the Freshman Introduction to Aerospace Engineering. Many around the world still go around in the fear that this will not work; we are now into our 21st year of Freshman Introduction and know that it works wonderfully well if taught by, well, people who know why our students come to engineering school. Your first assignment at the end of this chapter asks you to reflect on that question. Every time I read those answers from my students it reinforces all the excitement that I too felt - and still feel.

Beyond the Introduction, of course there is an infinite realm of books, papers, articles, art, and even music about aerospace engineering. People found the idea of Jet Travel absolutely fascinating. Consider the song from Peter, Paul and Mary, 1967 that started [1]:

I'm leaving.. on a jet plane
Don't know when I'll be back again..

or the one from the Steve Miller Band, 1977 [2] that went:

Big Ol' jet airliner
Don't carry me too far away..

My favorite is the beautiful, haunting 1977 lyric from Cosby, Still and Nash of the same title as the first line:

Just a song before I go
A lesson to be learned...
Traveling twice the speed of sound
It's easy to be burned!

Quite profound, and technically accurate. At ground level, where the temperature could be as high as 300 K (Kelvins, the absolute scale of temperature where water freezes at 273K), the stagnation temperature encountered, say at the nose of a vehicle traveling at Mach 2 is easily calculated, as you will see later, to be

$$T_0 = T + 0.2 * Mach^2 \tag{1.1}$$

That's 1.8 times 300, or 540 K as you can compute in your brain easily. Close to 270 degrees on the Celsius scale, you'll certainly get burns if you touch something that hot.

Science fiction books are very popular, and present many dreams of Humanity about flying around the Universe at unlimited Warp Speed and in supremely spacious comfort, luxury and amazing technical advancement, though in pursuit of very Medieval aspirations of killing each other or keeping from getting eaten. Many have become movies, adding to the large host of movies on the role of flying machines in travel and in war. People flock to see movies on say, missions to Mars, at least as much to criticize the technical realism or lack thereof in the movies as to watch the spectacular special effects and dream of playing the roles of the heroic actors. Just as an aside, yes, I have read some of the writing by Isaac Asimov [3], which was far from being fiction - those essays were true eye-openers on how to look at fields of science about which I was terrified before.

1.3   The Best in Action

I have read a very few science fiction books, but then I discovered books on the resources available [4] from extraterrestrial locations, meaning beyond Earth. I discovered the NASA Institute of Advanced Concepts (the original one, that existed from about 1997 until 2006) and that was truly a mind-blowing experience. I got to see real visionaries whose confidence in using any field of endeavor was truly inspiring. People doing research on how to build shelters by inflating or spraying on plastic balloons inside lava tubes and caves on other planets - and comparing those to the caves into which they routinely ventured (see Leaping Off Cliffs mentioned before.). People calmly discussing Quantum Teleportation (the idea that means, in its extremes, that you can disappear from here and appear a billion miles away at the same instant, no worries about the speed of light being a limiter!). There is a book by Michael Crichton [5] on many things that could go horribly wrong in that enterprise! Matter-antimatter propulsion. Did you know that one Absolute Limit cited by propulsion dreamers is the impulse that can be generated if a rocket converts a mass equivalent to Planet Earth into a flash of light energy by getting it to meet up with Anti-Planet Earth? Terraforming, building perfectly comfortable cities on Mars (why not the Sahara or Antarctica?), or floating in Space (my own preference, and our NIAC projects showed how to build those). People talking about hollowing out asteroids to build large spacecraft. Veterans of the Apollo program, whose confidence and knowledge in all disciplines was so supreme that they could grasp a presentation in any exotic field, and then type up a spreadsheet (I guess!) on their laptop computers, and come out with projections for how long it would take to build those, and make their business plans turn a profit! Suddenly, there was far too much to explore by myself, so there was no time and no patience to read wild science fiction. This will probably happen to you (I hope it does) as you learn the skills that we introduce in this book.

As we go through this book, we will introduce a very few technical papers, but more textbooks on the different 'disciplines'. I once got an email asking me, on behalf of a 'Fourth Grader', how a helicopter flies. Now explaining that to a PhD is easy, they'll nod their head as if they understand and go away completely confused but parrot what I tell them, Greek letters, equations and all. Explaining to a 4th grader is immensely more difficult, and it took me half a day to craft that answer. I also made the mistake of guiding them to a web page put up by NASA. A very nice thank-you note came from the Big Bro or Big Sis who had emailed me. They said the 4th grader genius had downloaded this latest Technical Report from NASA and was using that - no need for my explanation!

No, we don't want to do that. Learn the simple stuff first, get the concepts into your mind, become comfortable with the common sense of it (sense that becomes common because you are quite sure of it, having got beaten up exploring all other options), before cluttering your brain with the intricate stuff that takes a few decades to properly understand - and find that it is only a hypothesis. Of course one cannot have any sense, common or otherwise, before one tries working through some questions and problems.

1.4   Books on Aerospace Engineering

There are several excellent textbooks set at the level of a college student learning about aerospace engineering. One is the textbook by Professor John D. Anderson [6]. Another book, set at a slightly higher level, but with excellent reference data, is the earlier book by Professor Shevell [7]. Dr. Tennekes, author of a famous and beautifully lucid textbook on the difficult science of Turbulence, has written a very elegant book introducing flying vehicles [8]. This book is remarkable for the simple links between flying vehicles ranging from insects to jumbo jets. It shows, for example, that a bird taking off for a flight across the Atlantic Ocean shares many design features with an airliner doing the same trip (nearly half its body weight is food). Like the Lindberghs [9], the bird waits for the wind to become strong enough and turns into it, so that the wind speed achieved in its takeoff run is enough to generate enough lift to support its weight. Simmons [10] presents the aerodynamics of model airplanes. Alexander and Vogel [11] tell us about birds, insects and other flyers in nature, and Alexander [12] compares biomechanical flight to larger aircraft. Brandt [13] uses an aircraft design perspective to introduce engineering. Damon [14] introduces the science of space flight. Kemp [15] discusses the opportunities opened by the advent of sub-orbital space flights. The Civil Air Patrol has published a nice introduction to flight [16]. Hill and Peterson [17] is an absolutely priceless book on Propulsion. Why do I say that? It was written back in 1970 or so, from the hard experience (common sense at a very deep level!) developed at the Massachusetts Institute of Technology's Gas Turbine laboratory. There have been many books that claim to be more modern, etc, but after extensive review, my colleagues appear to always come back to this book as the one to have, if one is to have just one book on propulsion. Maybe that's because we got good grades learning from it as undergrads ourselves, I do not know, but if anyone could keep me interested for more than a page as an undergrad and help me get good grades in some pretty scary classes, that's an awesome credential right there. The more I learn about actually trying to write a book, the more I am in awe of authors such as these (and many others). This book is not 'easy'! Other books on propulsion include an early textbook recognizing the existence of 'aerospace' applications by Hesse [18] several by Oates [19, 20], Murthy [21], Mattingly [22], Cumptsy [23], El Sayed [24], Farokhi [25], Heiser on hypersonic propulsion [26], Spang on the control of jet engines [27], Brown on spacecraft propulsion [28]. In the field of fluid dynamics and aerodynamics, there are many, many more books, I will try to list just a very few [10–12, 29–39]. I wish I could say that a later book is usually better, but this is often just not the case. Authors get pressure from publishers to make books more attractive to readers (translation by a cranky old professor: 'watered down') so one often cannot trace the answer to the question of WHY, very deeply in more recent books. Many intricate and beautiful methods have been dropped as being 'too hard' and replaced with superficial equations and graphs, or the catch-all advice to go to Computational Methods. In Structures I have few to list [40–48], though there is a universe out there - Structures is the field that actually teaches us how to build flight vehicles and keep them from falling apart, while keeping them incredibly light. The books by Peery - the recent one [40] and much older ones, are probably what most of your professors or grand-professors used when they were students. I included a few books on Dynamics in the list above. Although Dynamics is not all Structural Dynamics, the subject is often taught by Structures professors.

Flight vehicle design is, of course, a highly specialized subject. Lawson's book on How Designers Think [49] and Ullman [50] and Siewierek [51] are examples of books on the general field of Design. One widely-used resource for aircraft design is the series of textbooks by Professor Jan Roskam [52]. Likewise, there are many books, with many more coming out every year in the field of spacecraft design and space mission analyses [53].

1.5   Back To Design-Centered Introduction

Aircraft designers introduce many variables to calculate the values of parameters to a high level of accuracy. We use a much simpler approach using the average values of these parameters. While less accurate, this approach gives useful results quickly, and illustrates the process of designing a flight vehicle. Creating a new design is, well.. creating something that does not exist. Once you see that the whole contraption sort-of hangs together and will fly, you can always keep refining the design, removing assumptions, tightening tolerances, and it should only get better. That is where all those years of engineering tools and in-depth courses that (I teach in upper-level classes and) you master as an Expert, will come in handy. But you have to have something that works in the first place before you can improve it.

This lesson struck me as an undergraduate in Aircraft Design in the late 1970s, when my classmates, armed with a new Electronic Calculator, used elaborate formulae from the textbooks to compute the design Structural Load Factor for an airplane down to 4 decimal precision, as 2.8753 or some such. Our professor, a veteran designer, smiled: "Oh, you mean 3! That's usually what it is." Which saved me because that was the guessed number in my notebook, having been too lazy to look up said formulae, and too poor to have an Electronic Calculator.

Yes, we intend to look at exotic concepts and go far beyond the systems seen flying today. To get a sneak preview, there is a growing amount of work on moving to a hydrogen economy. An example is the book by Rifkin [54]. Hydrogen is routinely used as a rocket fuel in large quantities. There is intense research towards generating hydrogen economically and without having to waste a lot shipping it, by using concentrated solar energy. Unlike petroleum and coal, hydrogen is fundamentally available anywhere there is water, for instance the coastline of the oceans and big lakes. When hydrogen reacts with oxygen, water is regenerated, so there is no danger of running out of hydrogen. So as time goes on, the cost of producing hydrogen should keep coming down, and its availability should keep on rising. If all else fails, Jupiter, Saturn and the Sun have essentially limitless amounts available as I hear, and so do other star systems.

1.6   Example: Why Do You Want To Be An Aerospace Engineer?

This assignment is to collect your initial ideas and aspirations regarding aerospace engineering, and to get you used to submitting assignments. Please write a 500-word essay on "Why I am in aerospace engineering". Please pay particular attention to what you aim to be doing after graduation, 10 years later, and 20 years later. Also, what you think is going to happen in aerospace engineering in that period (what projects, advances, setbacks..) You can use Microsoft Word to count the number of words (see under "Tools" for "word count"). The length must be between 490 and 510 words. Please put in the actual number of words at the end of the essay. The other use of this assignment is that several scholarship contests ask for essays on similar topics, so this should be a good start.

Chapter 2

From Dreams To Design

Aerospace Engineering is about converting dreams to reality using science, engineering, imagination and determination. So we will first summarize today's dreams, and consider a route map of disciplines through which they can be converted to reality.

2.1 Today's Dreams

Although today's airplanes and spacecraft are amazing feats of technology, the field of aerospace engineering is still very young. There are so many immense challenges before us. Let us consider a few. As we consider each dream, we will have to wake up, and think a little about what exactly each entails. This is the first step in defining our requirements, putting numbers to the dreams.

Fly like a bird : Wouldn't it be nice if we could fly like birds? What exactly does that mean, though? We would have to fly in the speed range somewhere between 0 and about 160 kilometers (100 miles) per hour, We must be able to fly high enough and far enough to cross mountains and rivers, but we must also be able to land anywhere safely and at a very low landing speed. To avoid hitting the ground hard, the landing speed must be down in the range of 5 feet per second. The other feat that birds routinely accomplish is that of landing precisely on a branch and holding on to it as they spread and then fold their wings, hardly even shaking the branch. **That** is a lot tougher.

Commute by Air: This is a dream of millions of people who find themselves sitting in traffic for hours every day. We wish we could just fly from our home garage to the office parking lot and then back to our garages. What does this imply in technical terms? There are already some "flying cars", invented since the 1970s. However, if these are to become as affordable as today's cars, enough of them would have to be produced and sold. So we must imagine a traffic management system that can accommodate, say, 1 million cars over a big city. These cars may be moving at 200 kilometers per hour, and they must be able to operate safely in all types of weather, including thunderstorms, wind gusts and blizzards, and at night. The main problem is collision avoidance,

11

keeping the commuters from killing themselves and each other. With self-driving cars and trucks actually plying the roads today, and autonomous Uninhabited Aerial Vehicles (UAVs) crossing the skies, we can anticipate that aerial commuting will include automatic piloting. Today in 2018, one crucial debate is about assigning responsibility for crashes: if the human occupant is not in control, who is responsible? The Air Traffic Control system (i.e., government) or the designers of the roads/airways, or the designers and manufacturers of the vehicles? Thus we see that critical decisions will be based on advances in public policy and jurisprudence, as much as in engineering.

A twist on the aerial commuter dream is a system aimed at the "early adopters" visualized as very wealthy visionaries who will pioneer these if it meets a dire need. The problem of commuting across the San Francisco Bay Area of northern California, USA, is often cited. Regardless of whether one drives a Chevrolet or a Lamborghini, a trip of 100 miles in workday traffic may take 5 hours! Work by NASA concluded that it would not be practical to have runways in residential neighborhoods. Besides land availability, regulations on noise during approach and takeoff would be prohibitive. On the other hand it was too difficult to make a flying car that could take off and land vertically, and yet function satisfactorily as a highway vehicle. One suggested solution was to design a VTOL (vertical takeoff and landing) Flying Car, that would take off from the billionaire's driveway, and land in the grassy circle within a highway clover-leaf interchange, where no one would notice the noise - and then hail a taxi for the rest of the trip! We thought this would be a bit crude, a billionaire arriving at a party in a mere taxi. Instead we proposed [55] an approach that would use a loitering Lighter Than Air (LTA) vehicle, and a slung cage under it. Only the cage would touch down, allowing the Lamborghini (or luxury SUV) to be driven in. The slung load would then be taken to the destination driveway or nearby, the vehicle would drive out of the cage, and arrive straight from the sky. Now is THAT stylish or what?

How many such LTAs could occupy the sky above an urban area at the same time? Probably fewer than the millions of cars on the roads, but more than the very small number that we currently imagine. Once the early adopters start making a concept visible, many alternatives and refinements will come up, and how the field evolves will probably be very different from how we imagine it today. Perhaps the present constraints on where people live in an urban area will change: for instance, the Bay Area, Seattle and many cities around the world, are severely constrained by the presence of waterways. People spend hours waiting for the ferry in Seattle, and due to bridge congestion in the Bay Area near San Francisco. If solutions such as the above are available, perhaps radically different solutions will emerge. This, incidentally, is the other lesson in Michael Crichton's book: how the collective intelligence of a 'swarm' will evolve, is not something that we can predict easily - even if the individual elements have little brain power or individual power. Imagine what millions of humans will think up - no, we cannot imagine, but we can be sure that it will be very interesting.

City to City, Doorstep Service: The next dream is to be able to fly from one city to another, across states, and again, be able to fly from our homes to our destinations. Such flight would have to be quite fast, perhaps at 400 to 600 kilometers per hour, so that we can visit friends and relatives who live, say, 1000 kilometers away. Perhaps in this case the vehicle will not be a personal one such as the commuter vehicle, but a larger craft that might pick us up from near our homes, like a bus. It would have to take off and land vertically or at least in a short distance, so that no long runway is needed. While landing and taking off vertically, it must not make much noise, or cause a strong downward blast of air.

Cross the World In a Day: Anyone who has travelled on a 14-hour or 17-hour airliner ride in economy class, will appreciate the desire for something that travels about 1.5 to 3 times as fast. Such a vehicle might cruise at Mach 1.4 to Mach 3, or 1.4 to 3 times the speed of sound. The speed of sound is roughly 600 miles or 960 kilometers per hour. Today, supersonic flight over populated areas is banned around the world: the sharp pressure waves can break windows and shatter eardrums. Researchers have steadily brought down the 'sonic boom overpressure' [56] below the acceptable value of 0.5 pounds per square foot, but that is done partly by flying at an altitude of 60,000 feet. Leaving pollutants such as carbon dioxide, oxides of nitrogen and even large amounts of water vapor at such altitudes is a dubious idea, because there are not enough regular winds or weather at such altitudes to blow or dissolve these away quickly, Perhaps we will argue later than a target of Mach 1.7 is much more feasible, and even 1.4 at 42,000 feet would be quite a valuable advance. With most people going around immersed in SmartPhones with headsets, perhaps few would even notice a sonic boom?

Note: Did you notice that we are being rather loose with our arithmetic? I didn't say that the speed of sound on a standard day in the Stratosphere above 33,000 feet altitude in the International Standard Atmosphere is precisely 583.47 Statute miles per hour etc. I have no idea if it is! That is quite all right for now. We need to develop the knack of making calculations in round numbers that we can work out while sitting on the bus to school or walking around the Student Center with headphones stuck on our ears. Once we get a grasp of the rough magnitudes of numbers, we can sit down with a calculator or computer and work things out in detailed and accurate mathematics. But simply being precise is no good if we don't have a grasp on the rough magnitudes. Professors get hammered by examples of such precision when we grade exams. Memorable examples include the dear student who cheerfully declared and underlined the answer of '10274.352873 meters per second' as his calculation of the Speed for Minimum Drag of NASA's Solar Pathfinder aircraft (I would expect it to be around 25 m/s ?) And the one who was shocked at not getting full credit for his answer of 33.2453 Newtons for the drag force acting on a 10-square meter surface of a hypersonic vehicle coming in at Mach 25 through the atmosphere.

Visit Low Earth Orbit: Many of us would like to experience the feeling of being in orbit around the Earth. The lowest altitudes where we can do this are about 300 to 400km above the surface. However, the issue in reaching orbit is not the height, but the speed. Going into orbit is like being a ball at the end of a string, being whirled around by a child. The string becomes taut as tangential speed of the ball increases, and the radial stress increases. If she lets go, the ball will zoom out along the tangent to the circle that she was describing in the air. But gravity does not let go, it keeps pulling us inward towards the center of the Earth. One has to be moving at a tangential speed of about 24,000 kilometers per hour, or around 7 kilometers per second (not the same but close enough!) to balance Earth's gravity and be in orbit.

Our Orbital Joyrider (at least for me!) should be a reusable spaceliner, and it must take off, accelerate, re-enter the atmosphere, and land, routinely and in absolute safety and comfort. The "National Aerospace Plane" project announced by President Reagan was promptly dubbed the

"Orient Express" by the media. As advertised, it was to be a hypersonic airliner that could take you from Los Angeles or New York to Tokyo at many times the speed of sound, zooming to the edge of Space.

Consider that in the days of the Space Shuttle, NASA invited the public to watch the spectacular liftoff from parking lots located at a safe distance. Following the liftoff, people would scramble to get in their cars and try to beat the rush out of the parking lot. Well.. before the average person could get their car out of the lot, they would be hearing the NASA commentator remark that the Shuttle was crossing Africa. So the prospect of a hypersonic airliner sounds very exciting indeed. But consider the acceleration and deceleration at the start and end of the journey. Accelerating at 3 times the acceleration due to gravity (3 Gs) or the vibration of a typical Space Shuttle climb are not very pleasant, even for highly trained astronauts. The airline would not have to feel bad about not serving any lunch on those expensive flights, as most passengers would not be able to eat any or keep it down for long. Not surprisingly, the National Aerospace Plane remains a dream, but surely some day some version of it will come about. You may be one of its designers who solves it's toughest problems.

Reaching orbit is like throwing a ball. Once the ball leaves your hand, it is in a ballistic trajectory (imposing-sounding word, but see the first 4 letters? Aerospace engineers used to be kids too, and the field is so exciting that it keeps us feeling young). The "throw" is the acceleration phase, when you are exerting thrust on the ball, and imparting an impulse or a packet of energy to the ball. If you think about what a space vehicle does, its engines "burn" for the first few minutes, and then it "coasts" for millions and millions of miles. We will see more on this when we study about spacecraft, but if you are burning fuel to produce this impulse, then, clearly, you want to burn up all the fuel as quickly as possible, because you don't want to spend the impulse on accelerating the fuel mass itself. So the ideal way to launch something is to put in all the impulse at the very start, as quickly as you can. Although a space launch rocket burn appears to go on for about an hour, this is still a very small part of the trajectory - much less than your arm swing before releasing a ball, as a fraction of the whole trajectory of the ball.

It still costs a huge amount of money to launch anything into Space. Very little of this is the actual cost of fuel, or even of the materials used to build the vehicle. Most of the cost is in the time and expertise it takes to design, fabricate, analyze, test, document, inspect, monitor and control the entire process. In other words, it is in the salaries of the people who work in the program. Without them, and without their full attention and expertise, space flight would not only be very difficult, it would be unacceptably dangerous. And no, this is not "government waste" and no, aerospace engineers do not get paid outrageous salaries or bonuses. But the net result is that today, the cost of sending anything to orbit, or to speeds similar to orbital speeds (as for that Saturday morning trip to Tokyo Mall) is immense. It costs something like $30,000 per kilogram of mass, to go to orbit. Yes, that means that if you weigh 180 lbs on Earth, your mass is around 82 kilograms. It will cost at least $2.46 million to send you to orbit - or to Tokyo Mall. For much less than that, you could go on the Internet from the comfort of your home, and in less time than it takes to say "10..9... Ignition!" you could charge that on your credit card and have the items shipped to you next day - usually on an airplane that aerospace engineers like you designed. This is the toughest problem of all in the Space program. The launch cost to orbit must be brought to about $100 per kilogram or less, to make it really worthwhile to do many things in Space.

Visit nearby planets: Wouldn't it be cool to go and prance around in the 0.3G gravity field of Mars, or see the frozen seas of Titan, or zoom through the rings of Saturn? Some day, surely, people will. Right now, the prospect requires that we get up to speeds of over 100,000 kmph; months of endurance because the distance is just so immense, and, most difficult of all, we need some way of shielding ourselves from the radiation in Space, once we are outside the protection afforded by the lovely blue skies of our atmosphere. The Sun, from where most of our energy comes, is one huge nuclear reactor, with both nuclear fission reactions (large, massive atoms breaking into smaller ones) and nuclear fusion (tiny atoms like hydrogen joining to form more massive ones) occurring. These processes give us sunlight and warmth, but also release powerful ultraviolet rays, gamma rays and X-rays, all of which can break the cells in our bodies and kill us. In addition to these, in Space there are other wandering high-energy particles running away from awful events such as exploding stars. These are called Cosmic Radiation and they may be particles traveling at maybe 75 kilometers per second. They can go right through most things, which sounds harmless, except that they may hit the aluminum or steel shell of our craft and cause thousands of smaller particles and rays to emanate from the inside of these shells and kill us. And on top of these, let's not forget those tiny specks of whatever floating about in orbit. They may be as small as dust, but they can punch a hole a couple of millimeters in diameter in metal skins or circuit boards - or through pressure suits and people. So look at those astronauts with a bit more respect. Yes, they ARE brave people, because for all the expertise of the Space Agencies looking after them, no one can really protect them against those last-named dangers. If you really are determined to stop radiation and microscopic particles, there is one way: surround yourself with about 2 meters (roughly 6 feet) thickness of soil, either from the Earth or, say, from the Moon. That will stop most things. Unfortunately, if you calculate the mass of a vehicle that has such thick walls, and multiply that by $30,000 per kilogram, you decide again that it's not time to take that trip. Some day someone will solve this problem. That someone may be you.

Visit nearby star systems: The nearest star to the Solar System is Proxima Centauri , which is about 4.24 light-years, or 40 trillion kilometers away. Obviously, to traverse such a distance in any acceptable time, a vehicle must travel at a speed that is a substantial fraction of the speed of light. This is the relativistic speed domain, where strange things are expected to occur.

Travel to Deep Space: Many interesting places including stars that have Earth-like planets, are millions of light years away. How we might start traveling to such places, is a matter of speculation. Our ideas of travel may be completely changed as we consider methods to travel to such places. Maybe *they* will tell us? Of course aerospace engineers have been busy considering this problem, and have broken it down to look at some of the implications, just as we will do in this book. It is clear that we will need propulsion systems that are far beyond the capabilities of the present. Some have tried to come up with absolute limits. One such limit is the energy that can be obtained if all the mass on Earth were converted to energy by interacting with an Anti-Earth comprised of Anti-Matter. But then one can argue that one could merely go and capture rocks that are floating around in Space and zap them with Anti-Asteroids instead of ourselves. Or zap Pluto or Planet X instead? Others have warned that any Anti-Matter Manufacturing (anti-manufacturing?) Facilities should be located well away from the Earth, perhaps close enough to the Sun that if things were to go wrong they would be pulled into the Sun by gravity quickly and gulped down in a flash. It is hard to imagine a facility on Earth that stores a great deal of the stuff: how will you contain it and make sure it does not come in contact with matter? No great reason for concern: the total amount of anti-matter generated so far is probably much less than 1 gram, and I hope all of that has been used up as people tried to make sure it WAS anti-matter by touching it to matter and watching the flash and applauding happily. If any has been saved up by people planning World Domination, I am equally happy to be unaware of it.

Reading the above will perhaps convey the wisdom of an important Design Law: We are not to go ahead with projects, where the total time needed for travel is more than 50 years. This is because (believe it or not!) *we expect to become twice as smart as we are now within 25 years*, so that we can design better systems and make the trip in the remaining time, much more effectively. This calculation was done before Social Media started taking up more and more of our time. Fifty years is a very short time in galactic terms. Perhaps the solution to interstellar travel will be much more exotic, such as accelerating to 'warp speed', entering a Different Dimension, and using a Space-Time Warp to magically wake up at the distant galaxy feeling refreshed and 'ready to roll' with the televised meeting with Their Leaders. Strangely enough, the idea of Quantum Teleportation, discussed elsewhere, suggests that the above scheme may not be any wilder. You could simply disappear from here and appear at the far-off galaxy simultaneously.

Nanoprobes: At the other end of the size and speed spectrum, is the quest to go ever smaller, and explore phenomena at tiny size levels. Nanoprobes are not necessarily as small as a nanometer, but may be built of pieces at the nanometer level (the size of molecules). Matter that is tailored at the molecular level can have vey exotic properties. For instance, one could imagine magnets that are millions of times more powerful per unit mass. What would happen to our design obstacles for electromagnetic propulsion if the magnets were super-light and super-powerful? Could aircraft simply levitate themselves off the parking lot before aerodynamic flight takes over?

2.2   Issues in Designing a Flight Vehicle

In Table 2.1, we summarize the process of flight vehicle design. The first column lists the steps, and the second says what each step does. The first step is to define the requirements, or decide what we want to do and why. That looks easy, but aircraft manufacturers will tell you that this is the most crucial step of all, and perhaps the one that is hardest to do as accurately as it must be done. This step requires the company to talk to hundreds, even thousands, of people, argue for hours, and eventually distill a clear idea of *what customers will pay, for the vehicle to do*. Why is this so? It is because the vehicle must eventually succeed in the marketplace, whether the customer is the military, a space agency, an airline, or a private owner. It must do better than predecessors have done, and yet not try to do too much.

2.2.1 Respect meets Optimism

The next step is, shockingly, to go and sneak a peek, or in fact spend a great deal of time and effort peeking, at what others have done before. This is called Benchmarking. Be aware that you should be careful how you obtain information: the penalties for industrial espionage are not a whole lot nicer than those for international espionage. But smart people like us, should be able to learn a great deal from legitimately published information, because we can empathize with the designers who went before us. What has been shown to be possible? For instance, has anyone managed to design an aircraft that can carry 2000 people and their pets and bags and enough food and water for the trip, non-stop at 700 miles an hour average speed, for 10,000 miles? If not, what exactly have they achieved? Why did they do that, and what stopped them from doing any better? You see that this gets pretty deep, but at the beginning it is as simple as listing the specifications of flight vehicles that are in the same general class of size, speed, capacity to carry useful load, and range, as the one that you are contemplating designing.

In this process, we adopt two attitudes. First, we *respect* the engineers who designed the vehicles that flew in the past, or are flying today, and assume that they did about as well as anyone could have done, and that they were very smart and very hard-working people. In other words, a very hard act to follow. Secondly, we *boldly assume* that we will also learn what they learned, and quicker, since we can learn from their experience. Given this advantage, we are expected to do better. So we can do better, but not immensely better, than they did. This is the culture of aerospace engineering. Brash we may sound, but inside we have a healthy respect for our competition, because if they are anywhere near as smart and quick as we are, well, they must be pretty awesome too. But we can't afford to sit around being over-awed by them, because we *have* to do better. I am sure that a modern aircraft company can also build, say, the Wright Flyer, a Messerchmitt Me-109, a Supermarine Spitfire, a MiG-15 or a Douglas DC-3, and they were all awesome technological marvels of their time, but will they succeed in today's marketplace?

2.2.2 Iteration and The Human Brain

The rest of the design process is pretty straightforward and logical. Please read the table, and as we get to each topic later, we will detail it. One observation to make: Note that the design process is *iterative*. We do not sit around until we can magically come out with a perfect design at the first attempt. Everything is linked to everything else, so where does one start? We break out of this dilemma by making a bold guess at the final answer, say how much the vehicle will weigh when it is all done. This is where it helps that we have the answers that our predecessors achieved. We will guess something a little better. Then figure out all the things that this implies, and suddenly we have a detailed specification of the vehicle. Maybe it won't work, and if that is the case, we will go back and refine our initial guess, and do everything over again, until everything fits perfectly.

In the old days this was painful, as one had to everything with a slide rule and pencil and eraser - but aerospace engineers still did it! Today it is so much easier to do with a spreadsheet on a computer. And take heart. My freshman students in 2010 complained bitterly at the immense

Step	Issues
Define Requirements	What must the vehicle do? Why?
Survey past designs	What has been shown to be possible?
Weight estimation	How much will it weigh, going by experience and our projection?
Aerodynamics	Wing size, speed, altitude, drag
Propulsion selection	Thrust needed? Number and weight of engines? Fuel consumption?
Performance	Fuel weight, take off distance, speed/altitude/ maneuvering boundaries
Configuration	How should it look? Designer's decisions needed!
Stability & Control	Locate & size the tail, flaps, elevators, ailerons etc. Fuel distribution.
Structure	Strength of each part, material, weight reduction, life prediction.
Detailed engineering	Design each part, see how to build, fit and maintain everything.
Life-cycle Cost	Minimize cost of owning the vehicle over its entire lifetime.
Iteration	Are all the assumptions satisfied? Refine the weight and the design.
Flight Simulation	Describe the vehicle using mathematics. Check the flight envelope.
Testing	Measure model characteristics, verify predictions. Build & test prototype.
Refinement	Reduce cost and complexity, improve performance, safety and reliability.

Table 2.1: Requirements

amount of calculations that they had to do in SIX weeks, to get their first design done. So much calculation! So many hours! Oh, their poor brains! Their hands as they had to type and write! They had never had to work so hard in high school, the poor dears! So I asked them to revise the entire design, for a vehicle that was quite different - it would use liquid hydrogen as fuel instead of the usual petroleum jet fuel. I gave them one week to do it. They did it. Then I asked them to do the essential parts of those calculations for another vehicle, as one question out of 6 on a 3-hour final examination, with no computer available to help them. So they had all of 30 minutes to do it. They complained that that was hard - but the point is, they did it, and did it very well. THAT, dear reader, is the power of iteration, when there is an aerospace engineer learning from the experience. So yes, most people find this stuff difficult. But persist, pay attention, and you will excel at it.

2.3   Vehicle Design: A Route Map of Disciplines

Aerospace Engineering involves many so-called *disciplines*: each might warrant a separate division in a major company, with dedicated experts who spend decades specializing in it. Here we take a quick look at some of these disciplines which you will encounter in this book. To become an expert in each of these disciplines, one should pay careful attention to the basic courses in school which don't always seem at first sight to be very relevant to aerospace engineering.

Much Remains to be Learned About Flight. Today's designs can fly over 100 times as fast as the Wright Flyer, and go right out into space, circle the earth every 90 minutes or so, and return to

Aspect	Basic disciplines needed
Mission Specification	Technology forecasting, market surveys, vehicle performance, economics, social sciences, political science
Weight Estimation	Statistics, technology forecasting
Aerodynamics	Physics, calculus, computer science
Propulsion	Physics, thermodynamics, chemistry, lasers, optics, environmental sciences, acoustics
Performance	Physics, Statics and Dynamics, calculus; flight mechanics
Structures	Materials, Statics, Dynamics, Strength of Materials.
Layout and detail design	Engineering Graphics, Psychology, Economics, Ergonomics
Stability	Statics, calculus
Controls	Laplace transforms, Differential equations, Electrical engineering, Computer science
Instrumentation & Communications	Optics, Electronics, Signal Processing, Computing
Space propulsion	Electricity, Magnetism, Nuclear Engineering, Chemistry, Physics, Dynamics, Thermodynamics
Trajectories & Space Mission Design	Dynamics, Astronomy, Modern Physics
Spacecraft design	Heat Transfer, Materials, Photoelectricity, Thermodynamics, Chemistry, Physics, Physiology, Electrical Circuits
Flight Simulation	Flight Mechanics, Image Processing, Engineering Graphics, Computer Science, Control Theory.
Ground and Flight Testing and Experimentation	Physics, Chemistry, Mechanical Design, Electronics, Signal Processing, Image Processing, Computer Science.
Lifecycle Cost	Manufacturing, Systems Engineering, Optimization, Economics, Political and Legal Issues.

Table 2.2: Route Map of Disciplines

precise touchdowns on earth. In the 1920s, a whole 17 years since the Wright Brothers made their first flight, the newspapers claimed that airplanes had reached the limits of speed and altitude. So-called Experts 'proved' that there was no 'scope' in this field. After all, they argued, what more could be done, than what had already been demonstrated?

So as of 2018, humans have around 115 years of powered flight experience. Birds and insects have 1 million-plus years of evolution, i.e., iterative design improvements, built into them. We cannot match the control precision, landing versatility, payload fraction, engine weight fraction, fuel costs, maneuverability, reconfigurable geometry, or structure weight fraction of birds and insects. By comparison with birds and insects, today's aircraft are still fragile and clumsy. They have stiff, nearly-rigid wings that can't flap, twist, fold or thrust to any significant degree. They need long runways and complex traffic control systems. You have to drive through 2 hours of traffic and spend another 2 hours at the airport and another 30 minutes on the taxiway to make a flight of 400 miles. When we launch spacecraft, only 10% to 15% of the total launch mass ever reaches orbit.

2.4   Summarizing

You are already well into aerospace engineering. You saw how the dreams of far-out concepts are immediately translated into tangible requirements with numbers. You confidently put down numbers from calculations done in your head, without worrying that the arithmetic was not exact to 5 decimal places. You have seen how this process gets us thinking about how to meet those requirements, and eventually tells us what is needed. You have seen that the issues and solutions come from all over the spectrum of human, and all other species' experience: we were not limited to merely 'aerodynamics' or other hard-core aerospace disciplines as we discussed everything from molecules and insects to Lamborghinis and radiation effects on humans. This is aerospace engineering! You have also seen some of the real obstacles in the way of achieving those dreams. Next, we will start the process of thinking through the requirements for something much closer to us: an airliner, or a "civilian transport aircraft".

2.5   Example Questions

Takeoff Mass: To do this assignment, you will have to hunt for information. Useful sources are your textbook, the worldwide web, and the university library. Find the maximum take-off weight in lbs, and the maximum length in feet, of the following vehicles:

1. Boeing 747

2. Space Shuttle (including boosters, external tank, and Orbiter)

3. Apollo 11 (Saturn V booster and everything inside and on top of it)

4. Lockheed VentureStar vehicle.

5. V-22 Osprey tiltrotor

6. Boeing-Sikorsky Comanche attack helicopter

7. Lockheed F-22 Raptor fighter aircraft

For each, please give the source where you found the information.

Example: Imagine The Realities Of A Dream Achieved: Choose one of the following topics, and write a short description (2 pages maximum; pictures are welcome). Use your imagination! A working day in the life of

- an astronaut assembling a Space Station in low Earth Orbit.

- the leader of an engineering team learning to operate a plant which produces hydrogen by splitting water, using solar electricity at the polar region of Mars.

- the Captain of a tourist spaceplane which will take 50 tourists to the Hotel Coweta in low Earth orbit.

- an Atlanta commuter who flies her "car" in from Athens to downtown Atlanta on a Friday morning in December 2010, coming in along the air corridor above what used to be the highways.

- the director of NASA Mission Control, on a day when a Mission carrying humans lifts off.

- the chief designer at XYZ Aerospace, the company which first decides to produce the first 1000 Supersonic Transport aircraft

Chapter 3

Estimating Take-Off Weight

Let us think about how an aircraft might be designed. How does one decide how big it should be, or its shape? How does one know whether it will travel as high and as fast and as far as we want it to do? All of its properties are related to one another. The designer needs a systematic process. If we knew how much the aircraft weighs, we could reason that it needs at least that much lift, and enough strength in the structure to support this weight, and so on. But until the aircraft is completely designed, how can one hope to know how much it weighs? Below we see how simple this process is. We start by guessing the most difficult answer of all: the Take Off Gross Weight of the aircraft. We try to reduce the uncertainty of this guess as much as possible by looking at what other people have done, and how those aircraft turned out in practice. There is nothing wrong in taking such a guess. We will systematically calculate everything else based on this guess, and then at the end we will see if the weight really turns out to be close to the guess. Otherwise, we will know how to change the guess and "converge" on the right answer after a few iterations of the procedure.

Requirements Definition / Mission Specification

First, we have to decide what we want the contraption to do. Then we will think of a "typical mission profile. To do these, we must find out what our customers really want, what others have been able to achieve in the past, and what opportunities are opened by new developments. In industry, this is a crucial stage requiring massive effort and intense thinking, because this is going to be the basis to commit the company's future. It involves discussions, analyses and trade studies with the airlines, financiers, regulating agencies, airports, law-makers and advertising or marketing agencies. Engineers have to learn to excel in this environment. Some of this process is decidedly 'fuzzy' in that we cannot imagine all the factors that will go into the decisions of human deciders. However, the Design community keeps trying to include everything that they learn, providing 'weighting factors' for each consideration. Decision-makers pay top dollar for 'data-based decision-making' optimization packages, and Big Data research tries to understand what is likely to happen: but the key point is that there are too many things that may occur or change in future. A hypothetical case is where a military agency is trying to decide whether an upcoming jet fighter aircraft should have one engine or two engines. Many arguments could be made on both sides. In one case, the agency gives this problem to a research team specializing in Multi Disciplinary Design Optimization. They give them a contract to come up with a data-driven decision in 6 months. Unfortunately the General in charge finds out that the answer is needed the very next week, for the program to continue to be funded. How is the decision made? We may never know.

Exercise: Mission Specification: Try thinking through this for yourself. No, we don't have a rigid answer. No one can prove you wrong, once you are 'in the ballpark' because there are no hydrogen-fueled airliners flying yet. We are giving one set of answers, perhaps you will come up with different ones. Are we right in insisting on independent operation with no jetway needed? When might this become really important?

Regional jet aircraft using hydrogen fuel
Range: 1600 km; Cruise Speed: 500 kmph minimum; 80 passengers

Typical Mission Requirements: Atlanta - Rayleigh-Durham NC, Albuquerque-San Antonio, Denver-Atlanta in summer
Short turn-around time. Fuel-efficient. Independent operation: No "jetway" needed. Reliable: All weather

Weight Estimation and Benchmarking

The mass to be carried is the "payload". This is the load which we get paid to carry. Once the payload is determined we ask, "haven't others tried to do something similar or close to this? How much did their aircraft weigh?" This is known as "benchmarking", getting a rough idea of the weight fractions of the various systems involved. For example, fuel weight may be 50% of the takeoff weight of a long-range airliner. In Table 3.1 we see how Takeoff Gross Weight (TOW) is Broken Out:

Note: However you break it out, you must make sure that everything is included somewhere, and only once.

Example: Computing TOW: Takeoff gross weight is simply the payload divided by the payload fraction. For example, if the payload is 30,000 lbs, and we find that *a reasonable payload fraction that we can achieve* is 0.15, then the TOW is 30,000/0.15 = 200,000 lbs. This is an estimate. You just learned how to get across the most difficult "canyons" of technical uncertainty in engineering: you jump across them. You make a "reasonable guess", see where it leads, see what is on the other side of the canyon, and then refine the guess as you learn more.

The rest of the design is to make sure that we come in under this estimate, when we calculate everything else. When we have a rough calculation of all the other things, we will go back and "iterate" many times to refine our estimates, so that the whole vehicle gets better. For this we

Component	Fraction
Payload Fraction: passengers+crew, baggage, food+water, cargo	$\frac{W_{payload}}{TOW}$
Propulsion Fraction: Engines, engine control systems, nacelles, fuel lines, fuel pumps, fuel tanks	$\frac{W_{engines}}{TOW}$
Structure and Controls Fraction: Everything else fixed to the aircraft: wings, fuselage, control surfaces, instruments, landing gear, hydraulic systems, air conditioning, lights, interior furnishings.	$\frac{W_{structure}}{TOW}$
Fuel Fraction	$\frac{W_{fuel}}{TOW}$
Total:	1.0

Table 3.1: How Takeoff Gross Weight Is Broken Out Into Fractions

will spare no technical effort, and it will take years.

Benchmarking

There is a wide range of answers to our question about the payload fraction. Some craft weigh only four times their payload; others ten times the payload. Payload fractions for aircraft which have similar missions and payloads should be similar - unless some radically new technology such as hydrogen fuel is used. In our case, one way of classifying missions may be passenger-miles, the product of the number of passengers and number of miles of range. There are many other ways of doing the classification.

Example Aircraft with Statistics

1. Embraer 190 Commercial Regional Jet, Brazil

Wingspan	23.56 m
Length	36.15 m
Height	10.48 m
Cruise Speed: Mach 0.80	

2. Boeing 787

Figure. 3.1: Boeing 787. Image courtesy of inslee.house.gov/issue/aerospace

Length	57 m
Wingspan	60 m
Wing Area	325 m^2
Height	16.9 m
Max Takeoff Weight	228,000 kg
Range	14,000 - 15,000 km
Cruise Speed	Mach 0.85

3. F-35 Joint Strike Fighter

Figure. 3.2: F35. Image courtesy of jsf.mil

Wingspan	35 feet
Wing Area	460 feet^2
Length	51 feet
Empty Weight	29,300 lb
Max Takeoff Weight	70,000 lb
Range	1,200 nmi

4. Sukhoi Su-30

Figure. 3.3: Sukhoi Su-30. Image Courtesy of the US Air Force. usaf.af.mil/shared/media/photodb/web/080713-F-4964T-02.JPG

Wingspan	14.70 m
Length	21.94 m
Wing Area	62 m^2
Height	6.36 m
Empty Mass	17,700 kg
Max Takeoff Mass	34,000 kg
Range	3,000 km

Design Step 2

6. Determine payload. Mass per passenger = ( ) kg. Supplies per passenger = ( ) kg. Baggage per passenger = ( ) kg. Cargo per flight = ( ) kg. Total payload = ( ) kg.

7. Determine payload mass fraction.

8. Determine Takeoff Gross Weight (TOW).

Example Problems

1. Find takeoff weights for several aircraft in the approximate range / payload class as your 300-seat intercontinental airliner. Try to determine their payload fractions, and any other weight fractions possible. This is to use as reference as you proceed with your design.

2. Please start a search for the history of hydrogen-powered aircraft. Specifically, our interest in seeing who has tried designing such aircraft for long-range passenger flight, and what were the problems encountered. You will not finish this job in half a week; the first step is a brief report on the progress of your search, and anything that you may have found already. In the 1970s, there was considerable interest, because hydrocarbon fuel was becoming very expensive. Lockheed definitely did studies on the drag reduction due to wings being cooled by the hydrogen inside. Ice formation was probably the killer of this idea.

3. Write a set of Specifications, and a typical Mission description, for the following market opportunity:

   At present, Overnight Mail delivery is done as follows (I think): The mail packets are collected and driven to the airport, loaded on aircraft, and sent to central points (like Memphis for Federal Express), From there they are loaded onto aircraft to get near the destination, then sorted and put on delivery trucks for given areas. The truck driver has to stop the truck at each house or office, get out, go to the door, ring the bell and wait a little, then (as occurs in most houses during the day) leave the package on the doorstep if allowed to do so, or leave a note saying "please pick up atÉ"

   Standard overnight envelopes are rectangular, stiff, and thin. To us AEs, they look rather promising as potential wings. Now if the driver could attach a small vehicle with an engine, propeller, tail, controls, a videocamera and a computer using a clip to the envelope, it should be possible to fly the envelope to the door, and either leave the envelope if the door opens, or if the driver says to leave it, or stick a sticky-note to the door and fly back. Obviously, we need a small wing to get the device back to the truck if the envelope is left, and we need enough range to come back with the envelope after sticking the note on the door.

   As you write the specifications and mission profile, remember that there may be another large market. Devices like these may be a blessing to Mail Carriers, enabling them to avoid being eaten by cute little Fifi, the 300-pound Dobermann. The Mail Carrier can sit in the safety of a HumVee and direct remote operations. The companies and Post Office may be willing to make some rule changes to accommodate these gadgets (the driver/ mail carrier rings the telephone by cell-phone control, rather than ring the doorbell, for example), but in general, it is wise to be able to do most of the duties in a manner similar to what the companies are used to doing.

4. Try estimating the market for such vehicles, including overnight carriers and mail carriers, so that we can get an idea of how much the development can cost, and how long it will take to break even, which in turn tells us how cheaply we can afford to sell these. Let that imagination fly again, as you did with the previous assignment. Note: The mission

specification includes answers to: "how high? how low? How fast? How long to stay in the air? How small? How much payload? The Profile is a typical flight description where most of the expected difficulties are encountered. Both must be as realistic as possible.

5. . Exercise your sketching skills. Sketch a picture (or multiple pictures if needed) of a bird coming in for a landing (wings outstretched, but not flapping), and mark (or describe if the part cannot be seen) the equivalents of the following aircraft parts:

   - wings
   - flaps
   - ailerons
   - tail
   - landing gear
   - fuselage
   - fuel storage
   - engines (note: the parts where the power is produced)
   - pilot/control system
   - control linkages (things that link the pilot's commands to the motors which actually move the control surfaces)
   - sensors for navigation

Chapter 4

Conceptual Design: Supersonic Airliner

Let us explore our newly-found skills, and go right ahead and design a 300-passenger supersonic airliner. Why not? At this stage, we are going to use several numbers from so-called experience. What follows is from a real discussion in our AE1601 class, embellished of course for this textbook.

It jumps way ahead in the sequence of presentation - and some may feel lost. Don't. Please have the patience to read through it. We will go back and cover the source of each number or argument presented here. Presenting this chapter so far near the front of the book is a way to reach some types of aerospace explorers, who want to see the whole panorama before them before they can have the patience to go down into the valley and wend their way across it. In addition, we want to convey that aerospace engineering is just not all "formulae" and "theory". It is based on solid human experience across the whole spectrum of human experience.

4.1 You must sell 500 planes to break even!

Developing a new aircraft is an immensely challenging proposition. The investment needed is not just in the aircraft itself: the companies that buy them must also invest heavily in many ways. Aircraft are usually bought on credit (a new one costs anywhere from 180 million to 450 million dollars today!) and every aircraft has to pay its own way by working hard night and day, staying idle on the ground as little as possible. Even at these prices, the manufacturer sets the price (watched closely by competitors!) as low as possible, assuming that some 500 units will be sold to spread out the extreme costs of developing the vehicle and the very very complex 'tooling' of the production line, and all that money spent on the glossy brochures and the slick advertising. The routes and the economics of passenger traffic must be carefully understood to keep the planes filled to capacity as much as possible, yet turn away as few passengers as possible - even if one were to forget the pain and inconvenience caused, every one becomes a potential gain for a competitor. Spare parts and trained crew must be stationed at many places for this to happen. Things can go horribly wrong, and insurance costs can be very high.

The 'gravy' in the airliner business is typically after the 500th is sold. The manufacturer has broken even at the Internal Rate of Return that they wanted. The product is clearly successful, the customers are happy. Other airlines and airports see the success of the model and want to emulate it. Orders come in for hundreds more, over a long period. Existing aircraft are brought back periodically for upgrades. The most successful airliners have 'sold' thousands. The massive and legendary Boeing 747, whose design started as the competing bid for the Cold War C-5 military airlifter, put the company at risk when that proposal was rejected. Boeing turned it into a civilian airliner. Sales crossed 1500 units in 2014, the first wide-body airliner (a.k.a. Jumbo Jet) to cross that milestone. The smaller Boeing 737 crossed a mind-boggling 10,000 orders in 2012.

The Anglo-French Concorde was perhaps the first demonstration that international consortium could develop a brave new concept. It was a brilliant technical success, but failed in the marketplace. What do I mean by that? Developing a commercial aircraft is a huge investment. Much of Boeing's brilliant success started with their early work in developing huge numbers of large bombers in World War 2. The manufacturing techniques and flight experience that they developed, helped them develop a series of commercial airliners, in parallel with aircraft for the military. The Douglas Aircraft company had early success with passenger-carrying aircraft, with the famous DC series. The DC-3 Dakota was built in immense numbers, and I had my first flight in one, all the way from Tiruchirappalli, Tamil Nadu, India, to Jaffna, Sri Lanka, in 1967. The flight took maybe 30 to 40 minutes, and I learned the cost of ignoring the flight attendant's kind advice to stick those cotton balls in my ears as I sat next to the big engines and propellers: I was left with a piercing ear-ache. The Douglas Skymaster was a successful international airliner. The DC-8 and DC-9 were highly successful jet airliners, I have enjoyed many flights in them. The DC-10, however, was not successful. Stories vary: some hold that the range was a bit shorter than what they had promised several airlines, and led to the cancellation of many orders. A couple of fatal crashes sealed the fate of the airliner - and perhaps of the whole company. The smaller DC-11 is perhaps quite similar to what became the Boeing 717.

To get back to what is needed for market success; in brief, one has to sell some 500 aircraft (OK, fewer if they are Airbus A-380s, maybe) to break even on the immense development cost and investment. Large airframe manufacturers claim that their net return on investment is a slim, slim 6 % per year. A large number of routes must become profitable to justify the purchase of 500 airplanes: no airline wants to keep planes idle on the ground. In fact, many airplanes seem to spend just a few hours on the ground between long flights, unless some serious maintenance is needed. And this required minimum number was the killer, when it came to the Concorde.

4.1.1   Why are there so few supersonic airliners?

The first routes for the Concorde were between coastal Europe and the coastal United States. The East Coast of the US was a viable supersonic destination: many businesspeople and wealthy socialites did the USA-'Continent' trip often. The promise of 'breakfast in London, lunch in New

York' seemed attractive. The other major route was Europe to Australia, stopping in India. Once that became viable, Australia-Japan might become attractive, and perhaps Japan- Hawaii, and then Hawaii to the US West Coast. Surely the Concorde marketing people thought of all these.

Flying at twice the speed of sound (the main advertised attraction: the interior was in fact pretty cramped!) the aircraft generated a loud Sonic Boom, the shock wave causing a very sharp N-shaped pressure fluctuation at ground level. I am not sure that I could tell a sonic boom from a thunderclap, except by looking out and seeing sunshine, but perhaps it would not be nice to have 6 of those waking me up through the night at equal intervals. People objected strongly to the idea of the Sonic Boom. Would large windows, ironically called French Windows, shatter? Would babies wake up and start bawling? Cows stop giving milk (a serious objection!) Crops wither? Soon, most of Europe banned overland supersonic flight. No problem! In those days planes that started in Western Europe could not fly over Eastern Europe or the Soviet Union or the People's Republic of China, so they could promptly head to the sea before going supersonic. But the United States banned overland supersonic commercial flight, and that was no doubt a killer. Neither US SST programs, nor the Concorde, could hope to enter the large US transcontinental market, nor offer a supersonic transcontinental route to points beyond. Australia followed suit. Seeing all these, India banned supersonic flight across the subcontinent. In those days the only economically viable (business center) airports in India were Mumbai, Delhi and Kolkatta, so this killed the plans for a refueling stop in India on the way to Singapore or Australia. Travel to South Africa was banned for citizens of most nations due to the Apartheid regime then in force. With this, there were no commercial routes left for supersonic flight, except London/Paris to New York, with perhaps a transonic extension to Washington DC. It did land elsewhere: I remember sitting in an airport restaurant at Hartford, Connecticut, and admiring the graceful Concorde taxiing there: evidently diverted by bad weather over New York, adding many hours of delay to that trip.

The US concepts by Lockheed and Boeing did not seem to get much beyond the development stage, though I have seen several reports and blueprints for those 300-passenger and Mach 3 aircraft designs. Perhaps the real focus was elsewhere: on the beautiful and deadly B-70 strategic bomber, designed to carry thermonuclear weapons at three times the speed of sound (one can be seen at the Air Force Museum in Dayton, Ohio).

By the way, the Soviet Union developed their own version of a supersonic airliner. The Soviet Tupolev Tu-144 shared several features of both the Concorde and the B-70. It was an ill-fated effort. They did attempt commercial service between the Far East and Moscow, flying supersonic over the vast steppes. But not for long: the plane was down with maintenance problems too often to be of much use as a fast and reliable conveyance. A fatal crash at the Paris Air Show shattered confidence in the program, and spelled the end of Soviet aspirations for supersonic airliners. Surely the design influenced the later development of the Tu-22 and Tu-160M supersonic bombers, some of which have seen successful combat operations in the Syrian War in 2016-17.

The first widely-announced flight of the Concorde carried media people and dignitaries from London to New York, arriving as promised, early enough for lunch. A minor problem: mysteriously, there was a well-organized, well-funded demonstration against the Sonic Boom and Airport Noise, choking the bridges connecting the airport to the city. It took passengers some 8 hours to make it into the business districts. Perhaps in time for a late dinner, but way too late to save the advertising campaign.

The Concorde hung on for decades, with practically the only routes being London-New York and Paris-New York. Occasionally it was used on charter flights, for example, to follow a solar eclipse, flying at the supersonic speed required to counter Earth's rotation speed (can you calculate that? Look up Earth's circumference: you have to go all round that in 24 hours). On its takeoff roll at Paris, a Concorde's tyres hit a piece of metal that had fallen off a preceding airplane. A tyre burst, pieces hit and slashed the wing fuel tank, and the plane caught fire just as it became airborne. The pilot tried to turn around to land, but the plane crashed well short of the runway, killing everyone on board. The remaining aircraft were grounded and the service cancelled; some are stationed at museums, including the Boeing museum at Seattle (where I was able to buy a ticket to take a trip through the Concorde for five dollars!) Oooh! Pretty cramped. Seats that look like (wonder why) car seats from the 1960s.

4.1.2   Revisting supersonics: the NASA High Speed Civil Transport

In the early 1990s, NASA initiated the HSCT program, partnering with the airframe makers Lockheed and Boeing, and the engine makers United Technologies and General Electric. Depending on whom you ask, the program was intended to produce a US supersonic transport, explore the design space between Mach 0 and 25, or just test out the new technologies of teleconferencing and video conferencing across the continental United States as an experiment on the Effects of Teleconferencing On The Stress Felt By Aerospace Engineering Managers. The program looked at the various aspects of high speed flight, including the prediction of trace oxides of nitrogen in high-temperature engines, the sonic boom, tyres for the high landing speeds, and new high-efficiency engines. Two designs were considered in greater detail. The first was a Mach 2.7 design to go faster than the Concorde. That encountered severe problems of landing speed. No tyres existed, that could survive the landings at the high landing speeds required by the design. Surface heating due to air friction was so high that after landing, it would be a while before the jetway could be brought to touch the hot aircraft surface! The final design considered was around Mach 1.7. This was roughly twice the speed of transonic cruise of most airliners, yet the sonic boom would not be too severe. Somewhere around that time, Boeing came out with a cool-looking design (aren't they all?) of the SoniCruiser, advertised as an airplane that could fly Mach 0.99. To those of us who had read the textbook chapters on Transonic Drag Rise, this seemed totally weird. Later explanation from senior Boeing engineers was that it was designed for Mach 1.4 (duh!) but they were not allowed to say that because Official Policy was that there was no plan for an SST.

In 1998 the HSCT program was cancelled! This was a huge disappointment to people like me who had looked forward to traveling on it by 2005 (OK, maybe the shortest flght, one way on super Clearance Sale, maybe on Friday the 13th, Christmas or Thanskgiving Day when the demand was lowest, returning by bus). In 2004, Boeing selected me to participate in their Welliver Summer Faculty Fellows program, one of about 160 university teachers so privileged, over a 17-year program. They wanted to teach us how the commercial world really worked, so that we could convey the message back to our students (and colleagues...). I spent the summer of 2004 wandering around Boeing, proud carrier of an ID card that let me into the 4th floor of their engineering building in Renton, Washington, where I could wander the aisles past the offices of all those PhDs at the top of the world's airline industry. They kindly informed me that Union rules 'prevented' me from doing any formal work. No complaint from me! Yes, yes, I could understand that they didn't want academics messing up their delicate instrumentation.

I asked everyone I could *WHY* they cancelled the HSCT. The answers were fascinating. Every discipline team (aerodynamics, acoustics, structure, propulsion) insisted that THEIR part worked fine, no problem. Dr. John McMasters, our Mentor at Boeing, spent hours talking to us, and he informed us that what we were seeing was the situation when **A Design Does Not 'Close'**. Which brings us to our lesson here.

4.1.3   Lesson: Revisit Market Surveys When Markets Change?

What did **NASA** say about why the HSCT was cancelled? Their Final Presentation was very clear on this. They said that supersonic commercial flight was a 'three-legged stool'. The three legs were Technology, Regulatory Approval, and Market Viability. They had clearly solved the technology issues. They were confident that could get regulatory approval (sonic boom intensity, and stratospheric pollution due to oxides of nitrogen). But.. the Market leg of the stool was broken.

Why was that? People who worked at Boeing, and at certain large airlines, confirmed this story: most airlines make much of their profits from the First Class and Business Class passengers, who pay 2 to 5 times what the economy class passenger pays, in order to get slightly larger seats, more swipes at the cookie baskets, heavy glasses for the Diet Coke instead of light plastic, and a higher probability of smiling faces rather than the usual. They were worried that if they bought supersonic airliners, those would simply sweep up all these Premier Customers, and their profits on the transonic operations would plunge. Why go for such a risky investment and shoot themselves in the foot, so to speak?

But clearly Boeing did not consider this to be a final argument. For one thing, they pointed out, the profit margin from building transonic airliners was already quite slim, and the competition was fierce. Transonic airliners were beginning to look all alike, and, like personal computers and cars, were fast moving towards becoming 'commodity items'. This was a far cry from the 1970s, when airliners were exotic marvels airlines invited people to Come Fly The Friendly Skies, etc. and Captains were one step above Emperors in my view. If someone else in some other country were to come in and offer to mass-produce transonic airliners, say, 20 percent cheaper, that would drive both Boeing and EADS (the parent company of the Airbus line of aircraft) out of the commercial aircraft business. So it was essential to keep moving ahead in technology, and a supersonic offering would surely be a high-ticket, specialized item, very hard for someone else to catch up and compete.

What caught my attention was the beautiful picture of the Globe that accompanied this 3-Legged Stool slide. It showed, of course, the Western Hemisphere, as NASA should show. It marked all the supersonic routes that were then talked about: US East Coast to Europe West Coast and Britain; US West Coast to Hawaii and Japan. That picture was unchanged from the 1960s Concorde days.

What was missing? Only the Eastern Hemisphere, and connections to South and Central America. Until about 1990, flying across the Soviet Union and the People's Republic of China was a bit risky. So routes over the North Pole were not very viable (besides, they worried that hydraulic fluid would freeze, regardless of what the Standard Atmosphere charts said about Mean Temperature). There were not many viable destinations or origins in China. In India there were perhaps 3 viable international airports with sufficient business interests within a 1-hour drive: Mumbai, New Delhi, maybe Chennai, and very few people other than movie stars and politicians could afford supersonic travel anyway. South Africa was isolated due to Apartheid policies: most countries' passports were stamped Not Valid For Travel To South Africa or Rhodesia.

Demographics of international travel are vastly different today. The market is immense compared to 1990. China is the world's biggest manufacturing destination, and at least tens of millions of Chinese people are aging, well-off and eager to see the world and visit their grandbabies across the world. India, likewise, has a huge population of aging people who saved all their money, are now retired, the real estate equity under their homes has zoomed by orders of magnitude, and the stock market has been more than kind to their savings. Their children work hard in faraway places, too busy to visit them, so they have to go over there to be with their grandkids. The same applies all over the world. Airport wait times and security problems add a lot of stress, uncertainty and fear to travel. If you were a young engineer or doctor working 80 hours a week somewhere, would you rather have your 65-year-old parents fly on those long, long transonic flights with long waits at strange airports just to save a few thousand dollars, and then become seriously ill or collapse? Or would you pay the premium fare to zip them over in supersonic comfort? So what would it take to develop a really good, profitable supersonic airliner where you don't have to be a dictator or movie star to afford the fare? We will look at this issue.

4.1.4   The Scare of Peak Oil

Oh, by the way, right around 2004, the price of oil shot up. People were predicting the imminent occurrence of Peak Oil, where the *rate of increase of production* of fossil petroleum products fell below the *rate of increase of demand*. From that point onwards, prices would rise out of sight. Consumption would actually come down as people switched to other fuels - and so, refinery operators would not see the point in building new refineries, or to invest in renovating existing ones. As refinery breakdowns became more frequent, with supply disruptions, prices would rise even higher, driving more customers away, and so forth. The days of fossil fuels seemed numbered. The implications for aircraft fuel were more dire: it turns out that aviation jet fuel was heavily subsidized by the cost of petrol products for automobiles. As demand for petrol (gasoline) for cars shrank, subsidized jet fuel would also disappear.

It is true that since 2008, as 'fracking' cranked up, the supplies of fossil fuels have shot up, demand has not kept up and so the prices have stayed relatively low. But around the world, concerns about Carbon Dioxide and Climate Change are driving moves to hybrid and electric cars, reducing demand for petroleum, and therefore, for subsidized aviation fuel.

Enter hydrogen. As difficult and expensive as it is to produce and store, liquid hydrogen is a fabulous fuel. When burned with air, it releases more than 3.8 times as much heat per unit mass as petroleum fuels. But... its density is still an absurdly low 70 kilograms per cubic meter, compared to 750 to 780 kg per cubic meter for petroleum fuels. Also, it must stay at about 6 degrees Kelvin, or 6 K above Absolute Zero - to stay liquid. Any other form of hydrogen (except Metallic Hydrogen, something too exotic to consider until we consider inter-stellar flight) has far too low a density to even consider - the tanks would be massive, and the pressures needed would require very heavy tanks. Tough problems, which is why we went into Aerospace Engineering.

Yawn! OK, ok, isn't it time to stop rambling and get to the point?

4.2   LH2 SST level 1 Conceptual Design, Fall 2017

Aircraft design is as much about these geopolitical events and demographics as it is about aerodynamics and structures. I will post the results of our Fall 2017 Conceptual Design activity below. We will see what is behind each step. The calculations were done on a Microsoft Excel spreadsheet. Many people do not realize that a spreadsheet is much more than a sheet with little squares with stuff printed inside them. It is in fact like a massive spread-out calculator. You can decide what you want to put in each cell. Some cells will have explanatory words or constant numbers, like the diameter of the Earth, the Universal Gas Constant etc. Those you type in directly. Others will have calculation formulae (starting with an = sign and referring to other Cells of the sheet, as well as all sorts of mathematical expressions), but only the results show. You change the values in a related cell, and all these results update instantly. Miraculous for those of us who grew up learning FORTRAN, the numbers can change all over, not just in the next and following lines. So you can do 'recursive' calculations. You can explore what-if scenarios, a fancy term which means: I can adjust numbers until I get the results that I want, although I will use all absolutely honest and accurate formulae. This last feature means that you can keep iterating on your design until you are happy or at least OK with all the results.

4.2.1   Specifications

First, the specifications. Supersonic flight is most suited to the Point-to-Point (P2P) airline architecture, rather than the Hub-and-Spoke architecture. Hub-and Spoke refers to the notion of having a few Hub airports. People fly from surrounding areas to the large Hub airport (think Chicago O'Hare, Atlanta Hartsfield-Jackson, LAX, Paris DeGaulle, Tokyo Narita, Mumbai Shivaji, Delhi Indira Gandhi). Large airliners carrying 400 to even 800 people fly between the Hubs, because that is is more economical. Point-to-Point means direct flight between long-distance destinations, which cuts out the connection, transfer and long wait at the Hub airport, and all its inevitable congestion. The catch is that this will only work if the P2P airports have enough of a business base nearby to keep a steady stream of passengers flying on aircraft that are large enough to be viable. Supersonic speed becomes a bit less attractive if one has to wait 3 hours at connecting airports.

Boeing articulated the Point-to-Point architecture in the early 2000s to advance the Boeing 787 Dreamliner. So based on the Dreamliner, we will set the number of passengers at 300, a level that is apparently suitable for travel between many cities that provide enough passengers to fly both for business and family reasons. Speed is important to avoid health problems of sitting for many hours, or having to stay awake at odd hours at strange airports across time zones. This may sound exciting to you at the age of a student, but is not very attractive to less-young folks (like the author). Blood flow slows down, the legs get swollen, knees may not work, heart problems may occur. For instance in my experience, if you arrived on a flight that was not Air France at Paris DeGaulle, they used to park the plane out far from the terminal, make you get out into the January cold and guaranteed rain at 7AM, stand around at gunpoint while a uniformed officer checked your papers, and get on a bus that went round and round the airport for 45 minutes to get to the terminal while you hung on to the straps for dear life (I can only imagine the plight of a 5-foot tall 80-year-old who could not sleep as well as I sleep on flights..) Then one had to race for the Gate to the next flight because 2 hours of the 3-hour transfer interval were gone, and the US flight required you to show up 1 hour ahead. And one would then get stopped by more Officers wielding machine guns and asked why one seemed to be nervous....

Come to think of it, before they invented The Economy Comfort fare (which I am too cheap to buy), I used to have enough Frequent Flyer Miles to be allowed to sit in the Emergency Exit seats where I can stretch my legs on the DXB-ATL or ATL-ICN 14 to 17-hour flights, I have seen numerous times, young people get up, walk down the aisle and collapse in a heap, so that the jump-seat facing me was occupied through much of the flight with such people being resuscitated by the flight crew. I don't think I would have wanted my parents to be subjected to all that if I could afford reasonable supersonic direct flight tickets for them.

Much of the advantage of speed may be negated if the trip is broken with a halt of several hours at a strange airport (see the wonderful Paris experience described above), so we go for the absolute maximum range: half the circumference of the Earth, which is roughly 10,000 miles or 16,000 kilometers. This is not as restrictive as it looks: we will load up with more high-value cargo, or not load up so much fuel, If we are taking shorter trips. Some airlines may reconfigure the airplane to cram more seats in, though we hope not. We provide room for 10 Metric Tons (10 MT or 10,000 kg mass) of cargo. Designing the cargo space is left for a much later exercise. What high-value, high-urgency cargo weighs that much? Flowers are very light. Documents for courier delivery perhaps? Farm-fresh produce? Designer cabbages?

We choose an altitude of 14,000 meters, or roughly 42,000 feet, the usual altitude of long-distance airliners. We do not have to rise to 60,000 feet because we think we can keep the Sonic Boom down, since we are not going to fly faster than Mach 1.7. These choices come from Boeing's Chief of High Speed Aerodynamics, and from NASA studies. Boeing technical papers of the late 1990s argue even to go down to Mach 1.4, but for us, that would stretch the time beyond comfort limits. Yes, by the same iterative techniques, we could explore designs for Mach 1.4, shorter range. The values in Table 4.1 are all numbers, no formulae used.

Specified Values	
Passengers	300
Cargo, kg	20000
Range, km	16000
Mach Number	1.7
Altitude, m	14000

Table 4.1: Long range SST specification

4.2.2 Assumptions

Next we make some assumptions, tabulated in Table 4.2. These are based on 'Benchmarking' although they may have started with what is politely known in the aerospace community as 'WAG'. We have refined some of these numbers after several iterations already. As we go along we will check whether these assumptions hold true or are violated. The wildest-looking number is the payload fraction of 0.33, meaning that the payload is 33 percent of the take-off weight!! Experienced Experts will roll their eyes. Smile at them: it will drive them mad. We started with around 25 percent and slowly pushed this up. Yes! With hydrogen fuel you will get such numbers.

The next is the engine weight fraction of 15 percent. This is long established in aerospace design, I see no reason why this will be violated.. yet. Down the line we see that with modern engines of the technology level of the US F-35 Joint Strike Fighter, this is more than easily met. The Structure fraction of 0.3 is generous for transonic airliners, which can make do with 0.27, but we need a large hydrogen tank, and we have to fly fast, so the structure may need to be stronger than usual. Once you add up the payload fraction of 33 percent, the engines at 15 percent and the structure at 30 percent, you get 78 percent, leaving a measly 22 percent for the fuel. For comparison, a long-range transonic airliner pushes 42 percent fuel weight fraction. Why can we even dream of this? Because hydrogen delivers 3.8 times more heat per unit mass as petroleum, peanut oil or other biofuel. And because we don't need to carry the excess weight of fuel we need even less fuel. We assume a Wing Loading (weight divided by wing planform area) in cruise, of around 4000 Pascals. This is an average: obviously it starts out higher when the fuel tanks are full, and ends up lower when the fuel is mostly gone. We can play with this number quite a lot as the level of detail increases.

Next we make an assumption (guess): what is the achievable Lift to Drag Ratio (L/D) in cruise? The NASA High Speed Commercial Transport claimed values up to 11. The Concorde had maybe 8. We will assume no more than 7 because our hydrogen tank causes a large volume which will cause a high Supersonic Volume Drag coefficient. The Thrust-specific fuel consumption of the engines is another technological assumption. This is the closest that aircraft engine designers will give you instead of 'gas mileage'. Distance means nothing to the engine. What can be predicted and measured is the weight of fuel burned, per unit thrust generated, per hour. This is the Thrust-Specific Fuel Consumption (rate). Note the units carefully: Weight of fuel, not mass, because we need something in force units to cancel out against thrust. The amount of fuel burned per unit time, depends on the amount of thrust generated. The unit of time used here is *one hour, not one second*. These values of TSFC are given along with the other specifications of

Assumed Values	
Payload fraction	0.33
Engine fraction	0.15
Structure Fraction	0.3
Balance is fuel fraction, the closing criterion	0.22
Wing loading, N/m^2	4000
Achievable L/D cruise	7
Engine TSFC (hydrocarbon), per hr	0.6
H2/HC heat /mass	3.8

Table 4.2: Long range SST: assumptions

an engine by the manufacturer, such as the Sea Level Static Thrust. We assume a TSFC of 0.6, which means 0.6 Newtons of hydrocarbon fuel consumed per Newton of thrust generated, per hour. Or it could be 0.6 lbs of fuel consumed per lb of thrust generated, per hour. We believe that this is again quite conservative. Modern high-bypass turbofan engines used for instance on the Boeing 777 or Airbus A340 and A380, achieve 0.35. However, we cannot use high bypass engines because these have large diameter. Large engine diameter means much more supersonic drag due to frontal area. But I think 0.6 is eminently achievable with engines that are generally scaled up a bit from modern fighter aircraft engines, and the bypass ratio can be higher for the SST than it is for a fighter engine.

Later we will convert this to hydrogen fuel (you use only 1/3.8 as much weight of fuel in that case, which means a TSFC of 0.1579 per hour). In fact, engines designed for hydrogen could offer higher operating temperatures, which means better efficiency. Against this comes the complexity of handling liquid hydrogen, pumping it to the very high pressure required to inject it into the engine, and the process of vaporization. Having bubbles bursting in the pump is no fun: it causes 'cavitation' that breaks the blades. Most materials become very brittle when exposed to liquid hydrogen, and many react with this extremely corrosive fluid. If left unattended for a while, as in those infamous taxiway delays that we read about, the liquid hydrogen must be allowed to evaporate, which of course means putting in more fuel. During flight with those high temperatures at the wing and fuselage surfaces due to friction, vaporization must be managed: the fuel must be consumed in the engines at least as fast as it vaporizes.

4.3   Derived Values

Hope you are with me so far. Now let us use those values, in Table 4.3. We argued in class how many crew were needed, for a long flight (the flight time is about 9 hours as seen below). Three pilots should be enough: at least one, maybe two, can be in the cockpit at all times while the third is on rest. Twelve flight attendants seems good, for 300 passengers with enough rest time. Average mass per passenger is set at 72 kilograms, which means a weight of 159 lbs. That is an average across all the passengers. Average baggage per passenger is 30 kilograms, equivalent to international business class allowance. Each crew member is allowed only 15 kg on average. We

Derived/decided Values	
Crew: 3 pilots, 12 flt. attendants	15
Mass/person, kg	72
Baggage/pax, kg	30
Baggage/crew	15
Food/water/person, kg	10
Payload,mass kg	55055

Table 4.3: Long range SST: derived values

plan on 10 kg of food and water per person. Passengers and crew need to eat, drink and use the restrooms - though on airplanes they use vacuum toilets rather than spend 2 gallons of water per flush; hence the scary "whoosh!". With these allowances, plus the 10 MT of cargo, the total payload mass comes to over 55 MT.

4.3.1   Does the design close?

Do you realize where we are? We assumed a payload fraction, and we now know the payload. So we can find the takeoff weight (TOW) of the supersonic hydrogen-fueled airliner! In Table 4.4 we put in the formulae to calculate such things. The TOW turns out to be around 1.63 million Newtons. Good time for a Sanity Check. The numbers on the Boeing 787 Dreamliner shows that to weigh around 2.5 million Newtons. Why are we so much lighter? Because we have a higher payload fraction, 33 percent compared to the 21 or 23 on the B787. So we are OK. Our 22 percent allowable fuel fraction works out to a maximum allowable fuel weight of just under 360,000 N.

Let us do a very rudimentary estimate of the fuel needed. Here we are simplifying things greatly. As we will see soon, the lift L must equal the weight, for the aircraft to fly straight, and level. The thrust must equal the drag D, to keep the aircraft moving in steady level flight. In reality, as the aircraft flies along, its weight comes down due to fuel usage. If the speed and altitude are fixed, the aircraft will reduce its lift coefficient, meaning reducing the angle of attack (the attitude becomes less nose-up). The lift to drag ratio may or may not remain steady. The thrust needed will come down as well, and so will the fuel consumption rate. All these are included in some form of the famous Breguet Range Equation, which we then integrate over the entire trip. In some cases, the airplane may gradually rise as fuel is burned up, and then this height is used to glide further as the airplane descends towards landing.

Since we are assumed not to know Integral Calculus, we will do this much smarter, as shown in Table 4.4. We will assume the L/D to remain constant at the value when half the fuel has been used up. (For those who are proud of their knowledge of the Breguet Range Equation and Integral Calculus, we are are thus taking everything outside the integral.) Given the speed of sound at 14,000 meters, we can calculate the speed corresponding to Mach 1.7. It comes out just over 500 meters per second, or half a kilometer every second. If we fly at that speed for 16,000 kilometers, it will take 32,000 seconds, or nearly 9 hours. Since we know the average

Result Values	
Payload, N	539539
Takeoff wt, N	1634967
Sanity check: B787, N	2489200
Max allowable fuel weight, N	359693
Speed of sound at 14km	295
Flight speed, m/s	502
Time in hours cruise	8.86
Median weight, cruise	1455120
Thrust needed, N	207874
Fuel weight HC, N	1104984
Fuel weight LH2, N	290785
Fuel weight fraction	0.178

Table 4.4: Results from the level 1 long range SST conceptual design exercise

weight of the aircraft during the trip, we know that the average lift is equal to that weight for straight and level steady flight. We assume that the lift to drag ratio remains constant as well, so we know the drag, and must compensate for that with thrust. Thus we know the thrust required. Given the thrust-specific fuel consumption, we can find the weight of fuel burned in the whole time, if it were hydrocarbon (HC) fuel. The weight of HC fuel needed would have been over 1.1 million Newtons, compared to the maximum allowance of 360,000 Newtons! So, just to remind ourselves, *this aircraft mission would be quite impossible with hydrocarbon fuel.* No surprise.

But what if the fuel is hydrogen? Since hydrogen releases 3.8 times as much heat per unit mass as HC fuel, the fuel weight needed is only 290,785N. Quite doable: the fuel weight fraction comes out to only 17.8 percent. But remember that this for the cruise segment alone. Taxiing, takeoff, climb and acceleration take up significant amounts of fuel, so we may be borderline on the fuel fraction. No worries: if it turns out to be too much, we will reduce the payload fraction a bit. If it is really this low, we can push up the payload fraction for the same payload, thus reducing the takeoff weight; this will further reduce the fuel needed. Other things will work in our favor. While climb and acceleration will require operating the engines at high thrust, the gliding descent can be done with the engine thrust cut way back.

4.3.2   Sanity Checks

We can and must check a few more things, in Table 4.5. The thrust of a jet engine decreases as altitude increases. Many engines have their best thrust while standing still at sea level, which is one reason why companies prefer to give their Sea Level Static Thrust values. The thrust decreases generally proportional to density. In addition, the thrust also varies with flight Mach number, so that for subsonic flight, one usually gets the result that the best thrust is at zero speed ('looks fast standing still' etc!) However, at supersonic speeds with fairly low bypass engines like the ones that we will have to use, the thrust actually rises to a peak at a supersonic Mach number.

Can your engines do what is needed?	Values
Density ratio at cruise altitude	0.1844
Takeoff thrust, assuming density scaling	1409126
Thrust to TOW ratio above 0.3?	0.861868609
Engine T/W achievable	7
Engine weight	201304
Engine weight fraction below 0.15?	0.1231

Table 4.5: **Check on engine takeoff thrust and weight fraction**

With suitable optimization, we will get engines that generate at least as much thrust at the cruise Mach number, as they do if they were to stand still at the same altitude. So we will ignore any thrust variation with Mach number here.

The atmospheric density at 14000 meters is 0.226753 kg/cubic meter, compared to 1.225 at sea level. The ratio is 0.1844, so the thrust at 14000 m altitude is only 0.1844 of what it would be at sea level standard conditions. In other words, you must have engines that generate 1/0.1844 that much thrust at sea level. Another point is that the cruise thrust can be only 80 percent of the maximum available thrust at the cruise altitude; we do not dare to keep running the engines at maximum thrust for 9 hours at a time, at least 19 hours every day. The poor things will die too soon! So these engines must have a lot of thrust at sea level. Far above the 30 percent of Takeoff Weight that we require as an initial rule. In fact the installed thrust will be 86 percent of the TOW, enough to send the plane zooming nearly straight up like a fighter plane. The engine's own thrust to weight ratio is a function of technology. For instance, the F-135 engine of the US F-35 Joint Strike Fighter is reputed to have a thrust-to-weight ratio of 12. We won't assume anything that grand, let us assume that the T/W of the engine is 7. This allows us to calculate the engine weight. It turns out to be 12.31 percent of the Takeoff Weight of the aircraft, close but still below the 15 percent allowance that we assumed.

4.3.3 Concluding Remarks

Everything looks good so far. We have an amazing airplane that can take 300 passengers and 10 MT of cargo at Mach 1.7, clear across to the other side of the Earth in under 9 hours, flying at 14,000 meters. The payload fraction is an astonishing 33 percent. The engines are powerful enough to make the airplane handle like a fighter off the runway. All this works only because of the hydrogen fuel - there is absolutely no hope of a viable long-range supersonic airliner that runs on hydrocarbon fuel.

Is this for real? I honestly don't know.

Don't you think the Big Boyz and Girlz of Boeing, Lockheed, EADS and Rolls Royce would have figured this out by now? Have their System Engineers not checked out this part of the Design

Optimization Space? But what do you see as wrong above? I see nothing wrong. If people sneer at this, ask them what assumption or specification they find to be unrealistic, and let us see what we can do about it. "My uncle whose girfriend is a top engineer at Bubba's Acme Airplane Company said there ain't no way this can happen!" is not a valid reason for concern. Nor, ahem, is "NASA has not said this". So far it looks like the Laws of Physics with good engineering will allow us to meet those specifications. Yes, it does take excellent engineering, as it did for the Concorde and the Tu-144 and the XB-70 and the SR-71. Why would we be interested otherwise?

In closing, let me convey the advice of our High-School Intern Miss Montana (not her real name) back in the mid-2000s. She was at her final presentation of the NASA Summer Internship in a large auditorium, facing an audience of other Interns, faculty, and many engineering PhDs as well as graduate and undergraduate students, and no doubt feeling one step away from fainting, though you wouldn't guess that from her polished presentation. Sure enough, a smart-aleck graduate student tried to impress the audience with his useless collection of factoids, asking how she proposed to deal with (a few of) the extreme practical complexities of her proposed Interplanetary Parcel Service (IPS). I was about 1 second away from launching nuclear missiles, figuratively speaking, at said smart-aleck, when Miss Montana, with an air of supreme authority, showed why she had won NASA selection to work at our lab, and that she had absorbed her lessons supremely well. She *smiled kindly*. *"See this $4B development budget? We will hire smart aerospace engineers to solve all those problems!"*

Well.. a postscript after a rough experience in our High Speed Aerodynamics class assignment in Spring 2018: I do know what is wrong: the lift to drag ratio of 7 is more than what we have been able to calculate for the minimum achievable drag of a large-enough aircraft of this type at these Mach numbers. See Miss Montana above: we need smart engineers to solve that. But there is a lot of leeway in our engine and structure assumptions, so there is no need for panic.

The next time you see this airliner, it may have changed a bit because of such excitement. We may reduce the number of passengers to 250, for one thing. This is design, and the iterative process that we have laid out allows us to refine the design as everything from market surveys of how many tickets we can sell, what routes are likely to be profitable, and which airlines want what, down to things such as supersonic drag calculations.

Chapter 5

Laws of Motion

5.1 Force Balance

We are back to analyzing the physics of flight. The wings (and horizontal tails and canards to some extent) support the weight of the whole aircraft. The rest of the aircraft just hangs from these lifting surfaces. Of course the wings and tails themselves have weight. On most aircraft, the wings contain most of the fuel. We can use Newton's Laws of Motion to calculate the acceleration of an aircraft, and thus decide how the forces on the aircraft must be balanced to make it go in a desired flight path.

5.2 Newton's Laws of Motion

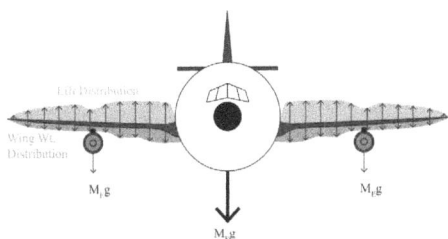

Figure. 5.1: Force Balance in Flight

The first law defines the concept of equilibrium. It says: An object continues to be in a state of rest or uniform motion unless there is a net force acting on it. So a state where there is no net force is a condition of equilibrium.

Newton's second law of motion states that the net force on an object is equal to the time rate of change of momentum. Let mass be m and speed be U. Then the magnitude of the momentum is mU. The rate of change with time is denoted by the differential calculus notation for the time derivative, $\frac{d()}{dt}$. Thus

$$F = \frac{d(mU)}{dt}$$

Using the chain rule of differentiation:

$$F = (\frac{dm}{dt})U + m(\frac{dU}{dt}) \tag{5.1}$$

If the mass remains unchanged,

$$F = m\frac{dU}{dt} = ma \tag{5.2}$$

In reality, force and acceleration are vectors: they have magnitude and direction. Mass is a scalar. The velocity can be denoted as

$$\vec{U} = u\vec{i} + v\vec{j} + w\vec{k} \tag{5.3}$$

where $\vec{i}$, $\vec{j}$, $\vec{k}$ are the unit vectors along the x, y and z coordinate directions respectively.

If two vectors are equal, i.e., $\vec{A} = \vec{B}$, then,

$$A_x\vec{i} + A_y\vec{j} + A_z\vec{k} = B_x\vec{i} + B_y\vec{j} + B_z\vec{z}$$

or

$$A_x = B_x, A_y = B_y, A_z = B_z$$

Using the above relation, the second law of motion can be written as 3 separate scalar equations, one along each of the 3 coordinate directions. We will use this fact as we describe aircraft motion in the next section.

Every action has an equal and opposite reaction. For example, if the engine of an airplane produces thrust which pulls forward on the aircraft, then the aircraft pulls on the engine in the opposite direction.

5.3   Describing Aircraft Motion: the Coordinate System

There is a book titled *'Equations of Motion: Adventure, Risk and Innovation'* [57] describing the life of the William Milliken, who did extensive work on the handling qualities of road vehicles. A revealing name, because the route to all the cool things we see in fancy maneuvers of cars and fighter planes, come from designs that start with the basic equations of motion. Since our interest is in the aircraft, and not so much on what happens to the air in a given place, we start by thinking of what we would see if we were attached to the aircraft. To learn this chapter, you do not need to know much about the properties of air. We merely refer to the forces of lift and drag. For the purposes of this chapter. these forces could come just as well from strings attached to the aircraft, as from aerodynamic forces. In a subsequent chapter you will learn about the properties of Earth's atmosphere and its air, and then learn about the science of aerodynamics. All in good time. For now, let us learn how to describe the motion of an aircraft using the laws of Physics, reduced to succinct equations that you can solve.

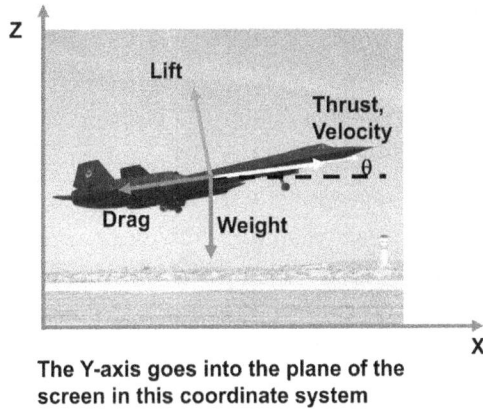

The Y-axis goes into the plane of the screen in this coordinate system

Figure. 5.2: Co-ordinate system: We usually use the Right-Hand Rule.

It appears that the aircraft is sitting still, and the whole atmosphere is rushing towards, past, and away from the craft. This is similar to the situation in a wind tunnel where aircraft models are tested. The model is held in place in the test section of the wind tunnel, and the wind speed is adjusted for each test condition. This leads to the concept of a *freestream speed*. The freestream is the air upstream of an aircraft. The freestream velocity is the velocity of this air relative to the aircraft. It is denoted as U_∞ or V_∞, though this text more commonly uses the former. An aircraft can be said to be moving at a velocity U_∞, though in that context it refers to the velocity with the same magnitude but opposite direction of the freestream's direction. The definition of the freestream is very important, for the definitions of lift and drag depend on it. Lift is defined as the force perpendicular to the freestream velocity (not as the upward or sideways force!), and drag is defined as the force parallel to the freestream velocity. We typically use a NED (north-east-down) coordinate system, with the x direction coming out of the the nose of the aircraft, the y direction exiting the starboard side of the aircraft. The z axis points down to the earth, and is positive down.

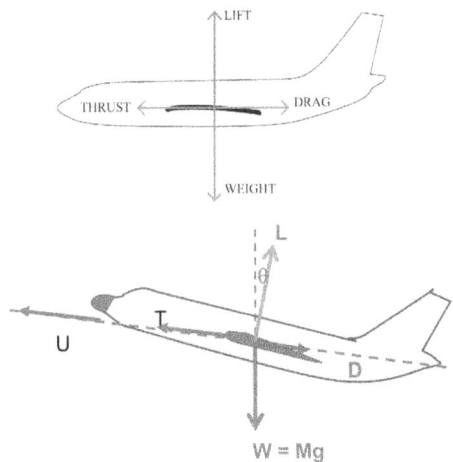

Figure. 5.3: Top: Steady Level Flight. Bottom: Steady climb with the angle θ small.

Weight acts towards the center of the earth (or whatever the closest massive heavenly body is). The two components of the vector equation describing Newton's Second Law (or Conservation of Momentum) are :

$$\text{Along x:} \quad -L\sin\theta - D\cos\theta + T\cos\theta = Ma_x$$

$$\text{Along z:} \quad L\cos\theta - D\sin\theta + T\sin\theta - W = Ma_z$$

The total acceleration vector of the aircraft is given by the vector sum of the components along the the x, y, and z directions. In the case above, there is no acceleration along the y-direction because there is no net force along that direction.

In **Straight and Level Flight**, straight and level steady flight, where all the accelerations are zero, lift is equal to weight, and thrust is equal to drag, with $L = W$ and $T = D$. In the case of **Steady Climb or Descent**, acceleration is zero, but $\theta \neq 0$.

A case of steady climb or descent will have constant speed but with the altitude gradually increasing. From the X-momentum equation, $L\sin\theta = (T - D)\cos\theta$
From the Z-momentum equation, $L\cos\theta + (T - D)\sin\theta = W$

From the X-momentum equation,

$$\tan\theta = (T-D)/L$$

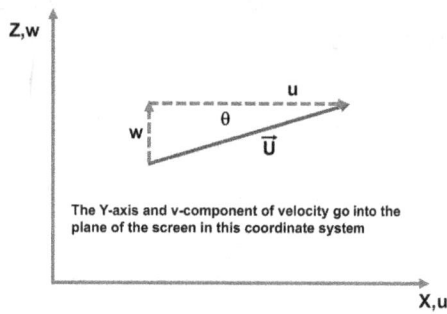

Figure. 5.4: Rate Of Climb

Let us denote the horizontal (X) component of velocity as u and the vertical component as w. From Figure 5.4, the ratio between the vertical and horizontal velocity components w and u is given by $\tan\theta = \frac{w}{u}$. Thus $w = u(T-D)/L$. Also, the relation between weight and lift is seen from the Z-component of the momentum equation above. If θ is small (as is usual under a routine climb condition where one is not in any desperate hurry), the value u is fairly close to the magnitude of the velocity vector, U. If w > 0, then the aircraft is climbing. From these we note, and experience confirms:

1) If the lift is greater than the weight, then the aircraft will accelerate upwards. Thus by accelerating in straight and level flight, the airplane can rise. Or, as the weight keeps going down with fuel being consumed, if the thrust and speed are kept the same, the airplane will rise. On the other hand, if the thrust is decreased, although the airplane may appear to be headed in level flight, it will descend. Once as a visitor to a major aircraft manufacturer, I was given a chance to sit in their Fighter simulator. I was flying at some 30,000 feet when these two enemy fighter planes appeared. I went after the first, coming in fast and high, then pulled back the thrust to get in behind the enemy and fired my missiles. Success!! Then I turned around towards the second enemy plane. I was so intent on following the target, that I failed to notice that since I pulled back the thrust, my airplane was descending. In the blue sky there are no cues. It was only when the brown earth came shooting up at me that I realized my error, and of course that was too late. No damage done, except that the rest of my team never allowed me to live that down for the rest of the trip. Or when they saw me years later.

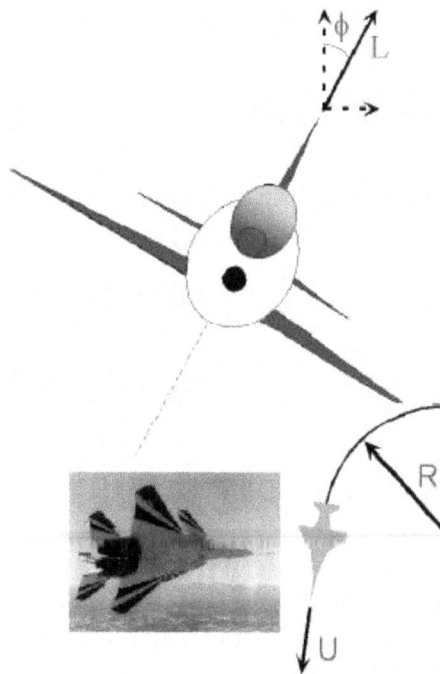

Figure. 5.5: Co-ordinated Turn

2) If the thrust is greater than the drag, the aircraft can climb if the thrust acts at an angle to the flight direction. Thus there are different ways of achieving the same result.

In the above, please note that keeping the airplane's nose, or more importantly its control surfaces, level, would counteract the climb or descent to some extent because the angle of attack would increase or decrease, opposite to its motion relative to the wind. In other words, if the airplane starts rising due to weight decreasing, the angle of attack will decrease because there is a downward component of the velocity of air relative to the aircraft. Thus the lift coefficient will decrease; the horizontal airspeed may increase until the aircraft is in equilibrium.

In some airline flights, the pilot sets the engines to operate at their most efficient thrust level in the vicinity of the desired speed and altitude, and then allows the airliner to climb slowly. The gain in potential energy can be used during the glide to the destination, by simply reducing the thrust. Thus some 30 minutes before landing you may hear the engine noise suddenly coming down and the flight getting quieter. I remember that on Boeing 727s (the engines were near the tail), one of my all-time favorite airplanes for business travel, in the days when seats were more comfortable and food service was something more than Peantus and Pretzels, things would get very quiet on the descent into Atlanta and I could sleep even better. Then the higher air disturbance at lower levels would rock me to sleep still better!

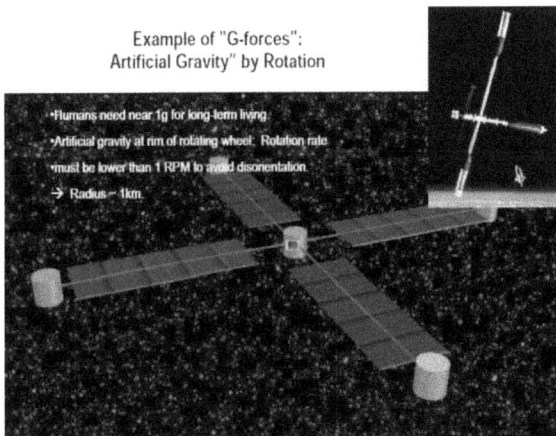

Figure. 5.6: A Space Station achieves artificial gravity by spinning.

Many years ago someone showed that endurance (the length of time that an airplane could stay aloft before the fuel ran too low and it was time to land) could be maximized by adopting the following scheme. The engines would be run at higher power at lower altitudes where there is more air per unit volume (higher density) and thus the engines were more efficient. This would cause the airplane to rise. Then the engines would be cut back to lower thrust and the airplane would eventually glide down, its speed being maintained because it was descending. For military airplanes, the changes in altitude in this mission could be quite large, but it is doubtful whether such a Porpoise Maneuver would be pleasant for passengers. Perhaps people like me would not mind at all but others might not be able to keep their food down. I cannot say if the engines would like this much either if done for most of the day and night, every day for 30 years as today's airliners work. The cost of doing more frequent engine repairs may outweigh any fuel savings.

Sideward Forces in a Level Turn can be found as follows: By rolling to an angle ϕ, the aircraft uses part of the lift force acting on the wings to execute turns, because lift acts perpendicular to the wings. As a result, a component of the lift, $L \sin \phi$, acts to make the aircraft turn. This force, called the centripetal force, is the force directed towards the center, and is equal to $\frac{MU^2}{R}$. The value $\frac{L \sin \phi}{M}$ is the radial acceleration. Note from Figure 5.5:

$$L \cos \phi = W$$

$$L \sin \phi = \frac{MU^2}{R}$$

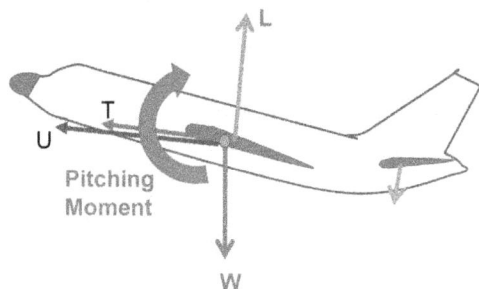

Figure. 5.7: The elevator is used to push the tail down so that the nose pitches up.

To pull tighter turns (smaller radius R) at a smaller value of U, $L\sin\phi$ must be made larger. If the aircraft is not to lose altitude during this maneuver, $L\cos\phi$ must be as large as the weight, W.

Artificial Gravity in Orbit The above reasoning also allows us to analyze what we need in orbit for **Artificial Gravity**. We see that we need a Centripetal Force to pull something around in a circle at any speed. Otherwise it will shoot off along the tangent. This is due to Inertia (see Newton's First Law about things preferring not to be disturbed and to keep doing whatever they were doing unless someone pushes/pulls/pokes them). However if you are in car being driven by an over-enthusiastic driver around a curve, you feel yourself sliding outward. You are already moving along the forward direction so you don't feel that. You only feel what you call the Centrifugal Force driving you outward. Some call it Inertia.

As a first year undergraduate, my Physics Professor had an aim with a piece of chalk was more accurate than a Sidewinder AIM9 Air-to-Air Missile if one dared to close one's eyes in the 1PM class in balmy tropical August in an auditorium: I've been hit more than once right on the forehead, to the accompaniment of merriment from my dear classmates.. Bless him, he tried very hard to teach me many things. In later years when I had no time, I dearly wished I had learned. But one exhortation that still rings between my ears is *THERE IS NO SUCH THING AS CENTRIFUGAL FORCE! It is only a REACTION TO THE CENTRIPETAL FORCE.* Oh, well! He was right of course, but all the same, Figure 5.6shows how one achieves Artificial Gravity in Space.

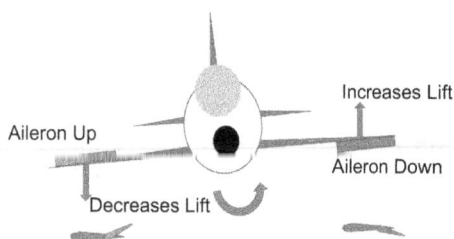

Figure. 5.8: Ailerons tilt in opposite directions, increasing lift on one wing while decreasing it on the other, thus making the aircraft roll

The picture at the top right corner is from NASA studies on how to use the Space Shuttle External Propellant Tank to build a Space Station with large capacity. The main picture is from our 1999-2004 study on Tailored Force Fields, done under the NASA Institute of Advanced Concepts, Phase 1 and Phase 2. Those 50m diameter habitat modules were to be formed from rubble brought from Near Earth Objects, into cylinder shapes with 2 meter thick walls, by electromagnetic force fields. Once formed by the force field, the rubble would be sintered into hard walls. One lives at the rim, and stands with one's feet pointed radially outward.

The inertia or whatever, makes one feel that one has weight. It is easy to calculate how fast the Space Station must spin, to achieve 1G (one Earth Surface Gravity, or an acceleration of 9.8 meters per second-squared. The central of hub portion of the Station is where one conducts one's Micro Gravity Dances. The residents in the modules at the periphery feel weight if they stand with their feel pointing outward. The above shows that if one wants to keep the spin rate below about 1 RPM (1 revolution per minute) in order to avoid upsetting the fluid accelerometer in one's inner ear, the radius needed is on the order of 1 kilometer! This is why it is not easy to provide 1G artificial gravity in Space. OK, perhaps half a G is pretty good. The Moon has only 1/6 G, and Mars about 1/3 G. Why should we care? Oh, because we are designed for 1G. Extended sojourns in microgravity can kill us. Modern astronauts try very hard to keep their muscles strong and their bones from getting weak, by exercising very strenuously, but still, many astronauts suffer from problems with blood circulation, large blood clots, and other mysterious ailments, even years after a long space-flight.

What happens in Roll, Pitch and Yaw? An airplane can use its control surfaces to pitch, roll, and yaw. Each of these is a rotation about the axes defined earlier. Pitch is typically controlled by the elevator, a horizontal control surface on the tail. Moving the elevator rotates the plane about the y-axis, which runs through the wings of the plane. As an aircraft takes off, it is tempting to just point the nose upwards and zoom into the Blue Yonder, but it is useful to remember that when one does that, one is actually pushing the tail down, and thus reducing the net lift on the aircraft, and Gravity is unforgiving in its pull downwards as well. If one does not wait until one is already floating well above the ground, one achieves what is called a Tail Strike: a surprisingly common occurrence even with large airliners.

Figure. 5.9: Rudder and Yaw

In the 1950s airplanes used to have a small tail wheel precisely to avoid disasters due to this. There was once a report that an airliner on takeoff from Heathrow airport, London, fully loaded for the long trip to China, experienced a tail strike: the Air Traffic Control people in the Tower (of the airport where they monitor aircraft, not London itself ..) saw the sparks/flash from the runway, and then the plane just kept on going... all the way to China. Not a recommended process: had there been a serious crack, it might have got bigger as the pressure difference between the fuselage and the outside got larger, and the tail may have broken off, taking the hydraulic lines for the elevator and the rudder with it, if not causing explosive decompression and (horror of horrors!) pulling the cart with the candy bars, cookies and the ice cream out the back! An airliner with no tail also becomes extremely hard to control, to put it mildly...

Roll is typically controlled by ailerons, horizontal control surfaces on the wings. Moving the ailerons rotates the airplane about the x-axis. **Yaw** is typically controlled by the rudder, the vertical control surface on the tail. Moving the rudder causes the plane to yaw, or rotate about the z-axis (vertical axis). Deflecting the rudder to the right causes the aircraft nose to yaw to the right. Note that in the original Wright Flyer, the rudder was out front so it worked opposite to what we just read, but the principle is the same. This is why it is a trick question to show someone the Wright Flyer without the pilot on it, and ask which way is forward to show how educated an aerospace engineer (if not historian) you are.

Parameter	Value in given units	SI or dimen- sionless
Altitude ceiling , feet	60000	18288
Cruise Mach Number		2
Supersonic cruise range, miles	1000	1600000
Return cruise Mach		0.7
Subsonic cruise altitude, feet	40000	12192
Vertical landing, unprep field		
Payload, lbs	10000	44545
Payload fraction at TOW		0.2
T/W, sea level at 80 percent of TOW		1.4
Single engine sl thrust, lbs	38000	169273

Table 5.1: Mission Specification

5.4   Example: Mission Specification of a Fighter Plane

The mission specification of a new fighter aircraft includes takeoff, climb to 60000 ft , cruise at Mach 2.0 for 1000 miles, loiter for 1 hour, engage in close combat at Mach 0.7 for 30 seconds, fire off 6 missiles, and return home with a supersonic dash at Mach 3.0 at 60,000 feet for 100 miles, followed by Mach 0.7 cruise at 40,000 feet and a vertical landing on an unprepared field. It is found that the payload (armaments) weighs 10,000 lbs, and that the payload fraction of Maximum Takeoff Weight is 0.2. Find the Max. T.O.W.

At 80 percent of the Max T.O.W., at sea-level, the thrust-to-weight ratio has to be at least 1.4 (the F-15s can accelerate going straight up; the F-22 thrust to weight ratio is given somewhere as 1: 0.72, or 1.39). The best engine suitable for use on this fighter has a maximum sea-level static thrust of 38,000 lbs. How many engines do you need on this aircraft? What is the takeoff thrust? What is the maximum acceleration assuming that the weight is at Max TOW at takeoff, if you take off and rotate so that the fighter is pointing straight up? If you divide this by $9.8 m/second^2$, you can express the acceleration in "G"s. How many "G"s does the pilot experience during this phase? (don't forget the 1G that she experiences all the time due to Earth).

Solution

Let us list the given information, and convert it to Standard International (SI) units of meters, kilograms, seconds, with forces in Newtons. I use a Microsoft Excel (or Apple Numbers) spreadsheet to do these calculations. You can enter a formula in each cell and copy that to other cells to repeat the calculation. For instance, in my spreadsheet, the weapon payload in pounds (lbs) is given in Cell C8, as 10,000. To convert this to Newtons in Cell D8, I use the formula: $= C8/2.2*9.8$. People use the term 'lbs' denoting 'pounds' to mean weight. One kilogram of mass has a weight of 2.2 lbs, or 9.8 Newtons (N). That calculates the result for payload as 44545 N. You can guess the formulae used for the other cells.

Max TOW is the payload divided by payload fraction		222727
80 percent of Max TOW		178182
Thrust needed to get the specified T/W		249455
# of engines needed, given single-engine thrust		1.47
Round-off to whole number of engines		2
Max thrust at sea level (sl) takeoff		338545
Vertical acceleration		5
Vertical acceleration, Gs		0.52
What the pilot feels in Gs, including the 1G felt in level flight		1.52

Table 5.2: Derived Results

The climb rate is an important specification for air superiority fighters and light interceptors. In air combat, height is a strong advantage, and she who gets to high altitude faster therefore has an advantage. The maximum climb rate exceeds 60,000 feet per minute but cannot be maintained all the way from takeoff, so it takes perhaps 1 to 2 minutes if one is seriously interested.

5.5   Further Reading

This very simple chapter is a drop in the vast ocean of Dynamics. A beautiful subject. If you begin to 'get the hang of it', you are well on your way to becoming a very successful engineer. Newton's laws of motion are used in aerospace problem solving very frequently. The fundamental concepts are as given above. Sir Isaac Newton's work titled Philosophiae Naturalis Principia Mathematica first appeared in 1687 in 4 volumes, written in Latin, as was the fashion in those days. An early English translation is [58]. There was research on Dynamics well before Newton's definitive work. A quick Google search under the term 'Engineering Mechanics: Dynamics' brought up dozens of textbooks - you can find them just as easily. A few examples are [59–64]. The more general field includes Statics and Dynamics [65,66]. Other interesting finds include [67–69]. To get more into philosophy, one might explore [70–72]. For those who might wish to leap right into Space-Time and Relativity, [73–78].

Chapter 6

The Atmosphere

6.1   Hydrostatic equation

The NASA Thesaurus [79] defines *atmosphere* as the mixture of gases surrounding the Earth, or filling the habitable volume of a spacecraft. The term is also used as a measure of pressure, meaning the pressure exerted by a column of mercury 760 mm high at 1 G $\left(9.8\frac{m}{s^2}\right)$, equal to 101,329 Pascals. Here we are discussing the defined concept first in the context of Earth, but generalized to atmospheres surrounding any celestial body.

Most textbooks on aerospace engineering and aerodynamics, give a table listing the properties of the International Standard Atmosphere, at intervals of altitude. Suggested examples include References [6, 7, 35, 37]. These days one can easily find Atmosphere Calculators on the Internet, so there is no need to dig out these tables. Below, you will learn enough to calculate the properties of air at any altitude from sea level up to 25,000 meters, with sufficient accuracy to not need either of the above props. If you want to know the properties in, say, Death Valley, CA or some such place that is below sea-level but not inundated, you can simply extend the ideas given here down from sea level. Above 25,000 meters things get less certain. We will give some ideas so that you can calculate there too.

The study of the different layers of the atmosphere is part of the field of aerostatics. This is a subset of hydrostatics, referring to the corresponding field where the medium is a liquid such as water, rather than a gas such as air. So what is aerostatics? Like hydrostatics, it deals mainly with a fluid at rest in the presence of gravity. Buoyancy, the net force that drives objects that are lighter than air or water upwards in the respective fluids, is an important part of aerostatics for us: balloons started flying a long time before heavier-than-air craft did.

A small digression into the story of Archimedes: many modern-day college students appear not to have heard this, so I am safe in embellishing it a bit. Dr. Archimedes was a historic figure [80–84] - the equivalent of Chief Scientist and Scientific Advisor to the King of his country. Famed for his brilliance and knowledge, though the Nobel Prize had not yet been invented, Archimedes enjoyed a revered, position and a comfortable home complete with a bathtub filled at the exact temperature of his command by his assistants. The King's new crown cost half the national treasure, and was gorgeous, but His Majesty suspected that he had been gypped as to its real gold content. He called in his Chief Scientist, and ordered him to find out the gold content of the crown -without scratching or otherwise damaging it.

Archimedes was stumped. As the standard 40-day period for such contracts drew to a close, and the Project Expiration Notices started coming out from the Office of Sponsored Programs (co-located with the National Dungeons), his terror and despondency rose. He knew all too well that the Head Scientist was well-named - his head was headed for the head-chopping block at the end of the 40 days.

On Day 40, Archimedes sadly went in to take his bath. As he sank in for one last comfortable bath, it occurred to his finely-honed senses, for the first time in his life, that he felt lighter than he did when sitting on a chair. He wondered desperately at the use of such an observation. As Ian Fleming said in a James Bond novel:

You only live twice.
Once when you are born,
and once when Death stares you in the face.

He was really living. And it hit him! *WHY* was he feeling lighter, the more he let his body sink into the water? And why was the water spilling over the rim of the tub? It was because he was displacing water. How much? Well.. as much as he had volume. And why was he feeling lighter? It was because this water was pushing him. Up. The neurons flashed. He knew how to calculate the volume of the water displaced, and that gave his own volume, as odd and pudgy as his shape was. He could do that for the crown as well, without scratching it one bit! And he could weigh the crown. So he knew its density! And he knew the density of pure gold.

Archimedes went wild with joy. He was going to live to see sunset. There was no Twitter, in fact no Internet. He leaped out of the tub and raced out into the streets, yelling: *I GOT IT!! I GOT IT!!*. Of course in his native language.

And what did the cynical Media of the day report? They whispered rumors that the Chief Scientist had been spied running down the street yelling incoherently about Getting something, in his birthday suit! The above is called Archimedes' Principle. An object immersed in a fluid feels a buoyant force equal to the weight of the fluid that it displaces. In other words, if that weight of displaced fluid is greater (and hence the upward force is greater) than the weight of the object itself, then the object is accelerated upwards like a balloon when you let go of the string tethering it to the ground. If the object is heavier, meaning that its mass per unit volume, or density, is greater than that of the fluid that its volume displaces, then it feels a reduction in weight equal to the difference between its own weight, and the weight of the fluid that it displaced. Archimedes did not have to measure the volume of liquid displaced by the crown, etc. All he had to do was to hang it on a spring balance as he dunked it in water, to measure the weight loss as percentage of the weight, and that told him the density. If that was not the known density of gold, the King had been cheated. Archimedes' principle is at the root of hydrostatics. It is at the root of Buoyancy. Ships use buoyancy. So do submarines. And balloons and airships. In some fashion it explains why the Ancient Temples etc are found buried under the relatively lighter (and looser and more fluid) soil around them.

Back to Earth's atmosphere. Earth's radius at the equator is about 6,378 km (3963 miles). The polar radius is about 6,357 km (3,950 miles). There are about 90 kilometers or 270,000 feet of gaseous atmosphere. Outer space is over a hundred miles up, but there is very little air above 51 miles. The Official Border between the atmosphere and Space is called the *von Karman Limit*, which is 100 kilometers. The significance of this is something like *Do not try to come to me with concepts for aerodynamic flight above 100 km*. In fact, spacecraft at re-entry to the atmosphere encounter maximum heating somewhere around this altitude, so there is certainly some air there! But serious aerodynamic forces start below this level.

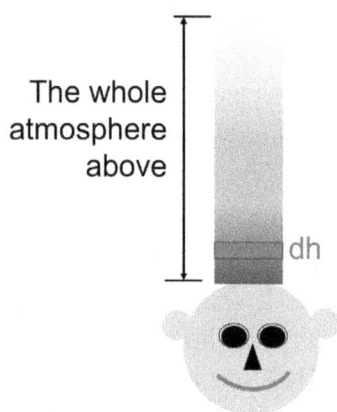

Because of gravity, the air above presses down on the air below. At sea level, air pressure is enough to support a column of mercury (Hg), 760 millimeters (mm) high: 101,325 N/m^2. For a given base area, this column of mercury weighs about the same as a column of air only 11 kilometers high at sea level air density. So most of the air is in the bottom layers of the atmosphere.

At a height h above the surface, let's say that pressure is p Newtons per square meter (N/m^2, or Pascals), and density is ρ kilograms per cubic meter (kg/m^3). The acceleration due to gravity is 9.8 meters per second squared (m/s^2). If you go up a tiny distance dh, the pressure decreases by a tiny amount dp.

$$dp = -\rho g dh$$

Figure. 6.1: *You support the weight of the atmosphere above.*

This is because you no longer have to support the weight of the element dh of the air column that went below you.

Perfect Gas Law

The Perfect Gas Law is a relation between pressure, density, temperature and composition of a gas.

$$p = \rho R T$$

R, the Gas Constant for the particular gas in question, depends only on the composition (i.e., the average molecular weight) of the gas, i.e., air. We know that air is generally composed of 20% diatomic oxygen (O_2; molecular weight MW = 32), 79% diatomic nitrogen (N_2; MW=28), and 1% argon (MW=44). Average (or "mean") molecular weight of air is given by

$$[(0.2)(32) + (0.79)(28) + (0.01)(44)] = 28.96$$

The Universal Gas Constant is 8314 J/kmol-K in SI units. Thus the gas constant for air is given by $R = (8314 J/kmol - K)/(28.96 kg/kmol)$, which gives a value of $R = 287.04 J/kg - K$.

6.2   Standard Atmosphere

People agreed to define a Standard Day, and the variation of atmospheric properties above Earth's surface on such a day. The first thing they decided was that the Standard Atmosphere would be defined at Mean Sea Level. The temperature at Mean Sea Level was defined as 288.12 K, which is just 15 degrees C or 59 degrees F. Why these you wonder? Why not 100 degrees F at 4000 feet above sea level? The answer is probably that this is what most pilots encountered when they went out for a flight on a peaceful morning somewhere around the East coast of the USA or around England in summer. But you can check into the history of how this was decided. It is not a big issue because we can use just about any Standard Atmosphere as long as it is defined well. The atmospheric pressure at Mean Sea Level accompanying this 288.12 K day, is 101,325 N/m^2, or 101,325 Pascals. The atmosphere's composition is such that the gas constant is 287.04, in the appropriate SI units (please figure that out for yourselves and check if that is right). The Perfect Gas Law gives a density of 1.225 kg/m^3. Please note these values, they come in very handy. Incidentally, one of the design specifications for a rotorcraft that can take off vertically, is the payload which it can hover, high enough above the ground to be out of ground effect, at 4000 feet above sea level, on a day when the air temperature is 100 degrees F. Thus **Sea-Level Standard Conditions** are: Temperature of 288.12 K, pressure of 101,325 N/m², and density of 1.225 kg/m³.

6.3   Regions of the Atmosphere

OK, we are ready to start calculating the atmosphere. Here another important simplification comes into the picture. This is that the atmosphere is divided into two kinds of regions. The first is an Isothermal Region: one where temperature does not change with altitude. The other is a Gradient Layer, where the temperature does change with altitude. The simplification is that in the gradient layer, the change in temperature is linear with altitude. In other words, if you know just one value for the gradient, for example, -6.5 deg. C per kilometer rise in altitude, and the temperature at some altitude in that layer, then you can can calculate the temperature at any altitude within that layer. It turns out that above the surface of Earth, we have a gradient layer called the Troposphere, extending all the way up to 11,000 meters altitude. Below 500 meters, we are in the Atmospheric Boundary Layer. The winds in this region get obstructed by hills, buildings, and by the friction of moving over the ground; hence they slow down and become turbulent. This is where we see most of the gusts, tornadoes, rain, snow, etc. Above this, and below 11,000 meters, is the Troposphere. Most of the "weather" occurs in this region, though some thunderstorms rise as high as 18,000 meters.

We mostly worry about just two regions: the Troposphere which is a Gradient Layer with a temperature gradient of -6.5K per 1000 meters rise in altitude, and the Stratosphere, which is an Isothermal Layer extending from the top of the Troposphere (11,000m, 216.65K) to 20,000m where it is still 216.65K.

6.3.1   Troposphere

In gradient regions, where T changes as altitude changes, we will assume that this variation is linear,

$$T_2 = T_1 + a(h_2 - h_1)$$

$$\frac{P_2}{P_1} = \left(\frac{T_2}{T_1}\right)^{\frac{-g}{aR}}$$

$$\frac{\rho_2}{\rho_1} = \left(\frac{T_2}{T_1}\right)^{\left(\frac{-g}{aR}-1\right)}$$

This holds in the Troposphere, the region between sea level and 11,000 meters. In the Troposphere, the constant a is approximately -0.0065 Kelvin per meter. Thus, for a standard sea-level temperature of 288.12 Kelvin, the temperature in the troposphere is given by:

$$T = 288.12 - 0.0065h$$

where h is in meters. In this region, the pressure and density variations can be found as follows:

$$\frac{P_2}{P_1} = \left(\frac{T_2}{T_1}\right)^{\frac{-g}{aR}}$$

density:

$$\frac{\rho_2}{\rho_1} = \left(\frac{T_2}{T_1}\right)^{\left[\frac{-g}{aR}-1\right]}$$

6.3.2   Stratosphere

From about 11,000 meters to 20,000 meters is the Stratosphere, where the temperature is constant at a cold 216.65 Kelvins. In other words, this is the *isothermal* region mentioned above. Most of today's airliners cruise in this region. Above that the temperature starts rising, again linearly, to around 265K at 45000m, remains constant to 50,000, and then things get more complex. Controlled airspace is only up to around 20,000 meters, so if you fly above that, well, you are on your own even more than you are beneath it. Composition starts changing approximately above 50,000 meters due to dissociation and ionization, caused by radiation and high-energy particles from space. In this region, differentiating the perfect gas law,

$$\frac{dp}{p} = \left(\frac{-g}{RT}\right)dh$$

$$\ln\left(\frac{P_2}{P_1}\right) = \left(\frac{-g}{RT}\right)(h_2 - h_1)$$

Since T is constant,

$$\frac{P_2}{P_1} = \frac{\rho_2}{\rho_1} = e^{\left(\frac{-g}{RT}\right)(h_2-h_1)}$$

Example: What is the standard temperature at 5000 meters? $T = 288.12 - (5000)(0.0065) = 255.62K$

We can express the pressure at a point on a given day as "so-many meters", meaning: "If I were in a Standard Atmosphere, and measured this pressure, I would be at this altitude". Similarly, we can express Density Altitude and Temperature Altitude. A major issue arises in tropical countries where there are high mountains, such as India. The Density Altitude is often much higher than Geometric Altitude. For example, in the Himalayas, where long runways are anyway hard to come by, landing speed becomes too high for the available field length on summer days. Helicopters find that they cannot clear ridges, and must fly through canyons and valleys instead, which is infinitely more dangerous.

Some Sample Values

Altitude (m)	Temp (K)	Density (kg/m^3)	Pressure (N/m^2)	Viscosity (N· s/m^2)
0	288.15	1.225	101,327	0.00001789
11,000 (end of troposphere)	216.50	0.363925	22,633	0.00001421
25,000 (end of stratosphere)	221.65	0.03946	2511.18	0.00001448
47,000 (end of linear temperature increase)	270.648	0.00142	110.9	0.00001703
60,000	245.45	0.00028	20.32	0.00001575
71,000	214.65	0.00006	3.96	0.00001410

Note, in summary:

1. It gets pretty cold and hard to breathe up there.

2. The "weather" is mostly below 11 km.

3. Most flight occurs below 20,000 meters today.

4. High-altitude winds can reach 200 mph.

5. The atmospheric boundary layer contains violent gusts and changes in conditions.

6.4   Other Planetary Atmospheres

The above description suggests that we can derive the atmospheric properties of other planets and moons. The presence of an atmosphere is a delicate balance between the gravity (and hence the mass and size) of the planet, the gravity of the nearby star or other large bodies (Jupiter and Saturn raid the atmospheres of their Moons periodically), and the heating of the planet by external and internal sources of energy. Where atmospheres have been detected, the composition varies widely. Venus is believed to have a hot, dense and mostly opaque atmosphere of carbon dioxide but is believed to be rich in Sulphur Dioxide. Mars has a thin atmosphere mostly of cold Carbon Dioxide: it is so cold that snowstorms of solid carbon dioxide are experienced. The moon Titan of Jupiter is believed to have an atmosphere (and ocean) of Methane, a compound formed from carbon, and the hydrogen which is the primary component of Jupiter's own atmosphere: Jupiter as a Gas Giant is all atmosphere, but down into its depths, the density may be quite high.

Seiff [85] compared data obtained by planetary entry probes, between Earth, Mars and Venus. The data on temperature and pressure profiles, and basic metereological conditions, were obtained as an instrumented craft descended, generally by parachute, through the atmospheres. These data suggest that linear fits to the troposphere data are reasonably accurate, so that a fairly simple model for these atmospheres is possible. This paper shows the complexity of the variations, and the approximations made to simply the variations. The data from these probes, obtained back in the 1960s, were used to design landers that came later. Data from the landers themselves complemented the cruder data obtained from the entry probes. Today, much is known about Mars, because numerous probes and several surface landers and rovers have operated there already. Much less is known about Venus, because the surface conditions are extremely hostile, and the atmosphere is no picnic either, having clouds of sulphuric acid, as one of many issues. The clouds are dense and there is a high (360kmph) wind circling the planet.

Let us put some numbers to those data. Mars' average surface pressure is around 600 Pascals (compared to 101325 at mean sea level for Earth). At the bottom of the deepest craters (Hellas Planita, 7152 meters deep) it reaches 1160 Pascals. At the peak of Olympus Mons (21,287 m above the mean level, the pressure reaches 30 Pascals. The composition is roughly (because different authorities give different numbers) 95.9 percent carbon dioxide, 2 percent Argon, 1.9 percent nitrogen, 0.14 percent diatomic oxygen, and 0.06 percent carbon monoxide. These numbers are from Wikipedia as of January 2017. Seasonal winds coming off the polar regions can reach 400 kilometers per hour. The lowest layer has significant amounts of dust, which helps absorb heat. The middle layer has strong winds described as a jetstream. The upper atmosphere, also called the thermosphere, has relatively high temperatures due to solar heating. Some segregation of different gases is reported in this region, The exosphere above 200 km is rather non-uniform. The surface average temperature is -60 C or 213K, close to the temperature of Earth's stratosphere which is 216.6K. The surface temperature varies widely by location and season from -125C (148K) at the poles in winter to +20C (293 K) at summer midday at the equator.

On Venus, the surface temperature is 740K. The surface pressure is 93 bar, or 93 times the mean sea-level atmospheric pressure of Earth! Again, carbon dioxide is 96.5 percent (not far from Mars!), nitrogen is 3.5 percent, Sulfur dioxide is 150 parts per million, Argon 70 ppm, water vapor 20 ppm, carbon monoxide 17 ppm, Helium 12 ppm and Neon 7 ppm, hydrogen chloride 0.1 to 0.6 ppm and hydrogen fluoride 0.001 to 0.005 ppm. Opaque clouds of sulfuric acid are seen in the upper atmosphere. The upper atmosphere spins with respect to the planet, circling the planet every 4 days, whereas the planet's own day due to its own spin takes 243 days. The upper atmosphere wind reaches 360km per hour, but near the surface the wind speed is barely

10 km/h. The pressure and density near the surface are so high that the carbon dioxide is in a supercritical state, more liquid-like than gas, so that it forms essentially an ocean.

The Venusian Troposphere extends to 62,000 meters. At 100 km above the surface, the temperature reaches -112C or 161K, and the pressure 2.7Pascals. From about 62km to 73 km, the temperature is nearly constant at around 230K. This is the layer of the upper cloud deck. Above this, the temperature decreases again.

Modeling the different layers of an atmosphere is fairly straightforward as we have shown here, once we know certain factors such as the temperature lapse rate, the composition, and the gravitational acceleration. The gravitational acceleration may be straightforward to find, from the body's orbital parameters (see the chapter on Space Missions for simple methods) which provide information on its mass, and the observed radius. The composition may be found by spectroscopy, though it is largely a mystery to me how astrophysicists accomplish that. I imagine that it has something to do with determining how starlight gets filtered through the atmosphere if one is able to observe that. Of course, solar system bodies have by now had probes sent by Earthlings visit them and send back data as wireless, radio frequency or other signals. Determining the atmospheric temperature lapse rate, which varies between layers, is not a trivial undertaking because many factors come into play. Radiant heating from the star (Sun) heats up the upper layers, and much of it gets transmitted to the surface so that it can heat all intermediate layers. Reflection and re-radiation from the surface heats up the lower layers and partly drives weather. The *albedo* or whiteness of the surface has much to do with this; for instance, glaciers and snow cover reflect far more than solar panels on rooftops, or black granite. Absorption of this heat by water vapor or droplets in the atmosphere complicates the derivation of the lapse rate. The changing composition in different atmospheric layers may change the mean specific heat and thus the temperature in different layers. Experimental measurements and detailed models may be used to predict these properties.

6.4.1 Further Reading

Planetary atmospheres are a subject of intense and fast-growing interest! The interest goes far beyond the Solar System to 'exoplanets', or extra-solar planets, meaning those orbiting stars other than the Sun. Since our closest neighbors within about 6 light-years do not seem to have any promising satellites, this usually means stars much further away, but mostly still in our Milky Way galaxy. More and more planets are being discovered (over 4000 when I last checked numbers for the Kepler mission circa 2015) as our telescopes and diagnostic methods improve. Early detection methods used the observed perturbations in the orbits of stars to conclude that there must be planets nearby. More recently, the light from a star is observed, and the transit of a planet across it is detected as a slight dimming of this light. The size of planets that can be detected so far away is coming down from the super-planet sizes of Jupiter and Saturn, down to the dimensions of Earth. We are far from being able to look into such planets and see their residents walking around. Only a very very tiny piece of the sky has yet been studied with these new methods. Some books and papers for further reading are listed below, there are thousands more for you to explore. Professor Strobel's Astronomy Notes [86] gives good background information at a starting college level. Renyu Hu's PhD thesis [87] laid out what to expect in planetary atmospheres, back in 1941 with atmospheric photochemistry and surface features. The great driver in those days, as now, is the potential to find biological signatures, signs of life. Kuiper Gerard [88] wrote about planetary and satellite atmospheres in 1950. Owen [89] wrote about the composition and early history of the Martian atmosphere. Seager [90] wrote about the physical processes in exoplanet atmospheres. Zurek [91] wrote about the dynamics and in [92] about comparative aspects of the Martian atmosphere.

6.5   Calculators and Applets

"Atmosphere Calculator " from Professor Ilan Kroo's Web Page at Stanford University, is based on the 1976 Standard Atmosphere, upto 71000 meters. It is available at http://aero.stanford.edu/StdAtm.htr . The ICAO Atmosphere Calculator is at http://www.aviation.ch/tools-atmosphere.asp The New-Byte Atmospheric Calculator is at http://www.newbyte.co.il/calc.html and the NASA Atmosmodeler Simulator is at http://www.grc.nasa.gov/WWW/K-12/airplane/atmosi.html

6.6   Example Problems

For this assignment, and for future assignments, you should learn how to enter a formula in Microsoft Excel (or whatever spreadsheet program you prefer). You should learn how to make repetitive calculations for various parameters, and then plot the results.

1. In the troposphere, the temperature on a standard day at a certain altitude is 250K.
   a. What is the altitude in meters?
   b. What is the pressure there, in $\frac{N}{m^2}$?
   c. What is the density there, in $\frac{kg}{m^3}$?

2. Find the composition of the atmosphere at ground level on Mars, (if you have to choose, choose the equator). Assuming that the temperature on a sunny day in summer is 80 deg. F, calculate a "standard atmosphere" Table for Mars, listing temperature, pressure, density and speed of sound for altitudes from ground level to 50,000 meters. Please express all quantities as fractions of their values at ground level, and then give the ground level values. We will assume here that on Mars, the temperature keeps decreasing with increasing altitude, with a lapse rate twice as rapid as that on Earth. The ratio of specific heats (gamma) for carbon dioxide is 1.13. We will assume here that on Mars, the temperature keeps decreasing with increasing altitude, with a lapse rate twice as rapid as that on Earth. If you find scientific data on the Mars atmosphere, it will be nice to compare with that, and see what phenomena we are capturing, and what we are missing. If you find such data, please plot the values in your atmosphere against the data. Also see if they give values for the "viscosity"; if so, please list that too.

3. In the troposphere, the temperature changes with altitude, with the lapse rate being 6.5 deg. K per 1000 meters. A balloonist rises from sea level to a point 5000 meters above sea level, what is the ratio of the pressure at B to the pressure at A?

4. In the stratosphere, an airplane climbs through an altitude change of 1000 meters. What is the percentage change in pressure? The temperature in the stratosphere is constant at 216.7K.

5. In the troposphere of a very Earth-Like Planet (ELP) named EZCalcica, the standard se-level temperature and pressure are 300K and 100,000 $\frac{N}{m^2}$ respectively. Atmospheric composition is 70% nitrogen and 30% oxygen, and the temperature decreases at 5 K per kilometer in the troposphere, up to 15000 meters. Find the temperature, pressure and density at 8000 meters on a standard day.

6. The atmosphere on a recently-discovered planet is a mixture of methane CH4 (60%) and ammonia NH3 (40%). Surface temperature on a summer day is 300K and the pressure is 200,000 $\frac{N}{m^2}$. Find the density.

7. **Example : Aerostatics** A balloon is perfectly spherical, with radius 5 m. The material of the balloon, plus the string used to seal and tie it, has a mass of 80 kilograms. The balloon is filled with helium (He, atomic weight 4) at a pressure of 101325 N/m2, and is tied to a chair in a room where the pressure is 101300 $\frac{N}{m^2}$. The chair plus its occupant have a combined mass of 120 kg. The temperature of the helium is the same as that of the air in the room, 288K. Calculate the force pushing the balloon upwards, and the acceleration of the balloon if the string is untied and released. Calculate the velocity vector and altitude reached at 1 second into the flight. (DO NOT TRY THIS AT HOME OR ANYWHERE ELSE!) Airline pilots on final approach to Los Angeles International Airport reported: *"Near midair collision with a guy in a lawnchair"* Upon landing, several LAPD officers assured him that he was not in Tahiti. He did get to meet several people, and entered internet lore as a candidate for that year's Darwin Award. AFAIK, no You-Tube video exists on this event.

8. **Example 2 : Aerostatics** A balloon is perfectly spherical, with radius 0.2m. The material of the balloon, plus the string used to seal and tie it, has a mass of 10 grams. The balloon is filled with helium (atomic weight 4) at a pressure of 101400 $\frac{N}{m^2}$, and is tied to a chair in a room where the pressure is 101300 $\frac{N}{m^2}$. The temperature of the helium is the same as that of the air in the room, 300K. Calculate the force pushing the balloon upwards, and the acceleration of the balloon if the string is untied and released.

9. **Example: Mid-Air Drama** Back home on Earth, an airplane cabin is pressurized to the same pressure as the Standard Atmosphere at 5000 feet (this is called a Pressure Altitude of 5000 feet). The air density is equal to the density in the Standard Atmosphere at 6000 feet (this is called a Density Altitude of 6000 feet). The pilot says in a calm, conversational tone: "Please fasten your seat belts: if you look out the left windows, you can see a tornado that has just formed". A passenger in Row 3, who is used to ignoring what other people say, stands up to stretch his legs, which have become numb from sitting in a cramped position while he was getting bloated from eating peanuts and Pretzels for 4 hours. The Head Flight

Attendant commands from the back of the plane: "**SIT DOWN!**" The distance from the flight attendant to the offending passenger is 240 feet. How long does the front of the sound wave from this yell take to reach the offending passenger's ears? Since smoking is banned on this flight, and the air recirculation system was designed by AEs, we can assume that the air inside is a pure 79% N2 and 21% O2. Assuming that the aircraft is flying at 350mph, and the tornado is one mile away, how long does the passenger have to sit down and grab the seatbelt after the yell reaches him? The airplane will go into a steep banking, diving right turn when it is 0.8 miles away from the tornado.

10. **Example: ExoPlanet Atmosphere** The atmosphere on the ExoPlanet CGX373 in the Andromeda Galaxy (I am sure that the citizens attending college there know it by some other name) is a mixture of 2 gases: Methane (CH4): 80% and Hydrogen Cyanide (HCN): 20% (Assuming that these things do not react violently at this temperature and pressure!) The atmospheric temperature at our assigned Visitor Landing Deck is 310 deg. K. The pressure is 101325 $\frac{N}{m^2}$, just like Earth on a Standard Day at sea-level. Find the density.

6.7   Design Step 3

Determine conditions at cruise altitude. (Cruise altitude for short-range aircraft is lower than that for long-range)

Calculate the standard atmospheric conditions at a selected altitude for your aircraft.

On a day when sea-level temperature is 35 C and pressure is 101,000 Pascals, find the actual altitude if the aircraft altimeter indicates 12000 m.

1. Construct a spreadsheet calculation where you can specify the altitude and get all the standard conditions: temperature, pressure, density. Validate at several points by comparing your values with those you get from an Atmosphere table in a textbook or against what you find from Atmosphere Calculators on the Internet. Calculate the percentage error in your values, defined as ((Your value minus the value from the reference), divided by the reference and multiplied by 100).You should get values that are very close, certainly within 5 percent and usually much closer.

2. Construct a spreadsheet where you can specify a pressure and temperature, and you can find the pressure altitude and density altitude.

Chapter 7

Aerodynamic Lift and Drag

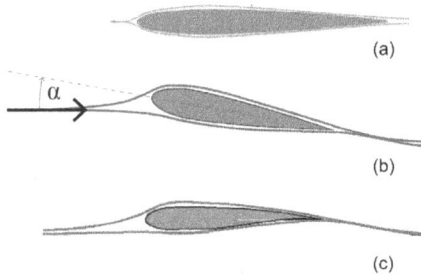

Figure. 7.1: Airfoil lift generation. (a) Symmetric airfoil at zero angle of attack produces zero lift. (b) symmetric airfoil at angle of attack. (c) Cambered airfoil.

Aerodynamics is the study of forces due to the movement of air around or through an object. It is a vast and somewhat complex subject comparing many different regimes where phenomena are quite different. Although rigorous analytical methods in aerodynamics use applied mathematics to a great extent, the final results turn out to be elegant and simple.

Aerodynamic lift is the force perpendicular to the freestream. It is generated by deflecting the freestream air. According to Newton's laws, lift is the reaction to the rate of change of momentum of air, perpendicular to the freestream. Lift is related to freestream speed by:

$$L = \frac{1}{2}\rho U_\infty^2 S C_L$$

where U_∞ is the freestream speed, C_L is the wing lift coefficient, and S is the planform area of the wing.

In low-speed flows of air (<0.3 times the speed of sound, or Mach 0.3), there are three main ways to create aerodynamic lift. All involve directing the momentum perpendicular to the freestream. The angle of attack can be varied, camber can be added to the wing, or lift can be induced through vortices.

Aerodynamic Drag is generally bad news, as the name indicates. It is the force along the freestream, which means the force dragging the airfoil back. It is due to irreversible loss of momentum. Drag occurs because of skin friction as the air flows around the airfoil. At low speeds this is mainly due to viscosity. As speed or airfoil size increase, the flow becomes turbulent, and loses some momentum just tumbling around. As angle of attack increases, the flow separates near the trailing edge. In the separated flow region, the air is pulled along with the airfoil, so the drag increases sharply.

Drag is given by:

$$D = \frac{1}{2}\rho U_\infty^2 S C_D$$

The lift to drag ratio is:

$$\frac{L}{D} = \frac{C_L}{C_D}$$

We want our planes to have as high a L/D ratio as possible! On the other hand, we want our parachutes and speed brakes (which some fighter planes have) to have high drag.

7.0.1   Varying the Angle of Attack

Imagine a wing with a symmetric airfoil at zero angle of attack as shown in Figure 7.1(a). The airstream around it is as shown. To be precise, the lines shown are called streamlines. At any point the local flow velocity vector is tangential to the streamline (a concept that applies only if the flow is steady). So from the streamline we can determine the local flow direction, assuming that we have no uncertainty about the sign (say to left or right).No air is deflected down in the case of the symmetric airfoil at zero angle of attack. Because the airfoil does not push any air down, there is no air pushing the wing up. A wing made of this airfoil creates no lift at this condition. Now see Part (b) of the figure. The nose is pushed up to positive angle of attack. Angle of attack is the angle between the chord, which is the line joining the leading and trailing edges, with the freestream. Positive is as shown. The streamlines are now deflected downward, which means that an amount of downward momentum is being added to the air every second. This rate of increase of downward momentum requires a force, which the airfoil is exerting on the air. In reaction, per Newton's Third Law of Motion, the air pushes right back at the airfoil, generating lift. Now go to Part (c) of the figure. Instead of tilting the leading edge and hence the chord upwards in positive angle of attack, we could simply curl the trailng edge downwards. The chord is still the line joining the leading and trailing edges. But now the *camber line* , which is the line that is equidistant from the upper and lower surfaces, is curved and no longer coincides with the chord. Even if the angle of attack is zero, the cambered airfoil generates lift. In fact one would have to tilt the airfoil down to some negative angle of attack before one could get zero lift. And sure enough, this angle is called the Zero Lift Angle of Attack: usually a negative number of degrees, like, say, -2 degrees.

Figure 7.2 shows how the lift coefficient of a symmetric airfoil varies with angle of attack, from 0 to 180 degrees. Since it is symmetric, the variation from 180 to 360 will be the mirror image of the variation from 0 to 180. The NACA0009 was developed by the National Advisory Council for Aeronautics, the predecessor of today's NASA. It uses a 4-digit numbering system. The first two digits give the amount of camber and the location of maximum camber. In this case the camber is zero. The last 2 digits give the ratio of the maximum thickness to the chord in percentage. Generally, the lift coefficient varies as a sinusoidal function, reaching its highest value at 45 degrees, touching zero at 90 degrees, and so forth. However, in the regime from about -12 to + 12 degrees, there is a strong, linear variation (the lift curve slope is constant), reaching a high value of lift coefficient around 12 degrees. Now let us look at the variation of the drag coefficient in Figure 7.3. The coefficient is small but finite at zero degrees, and then varies

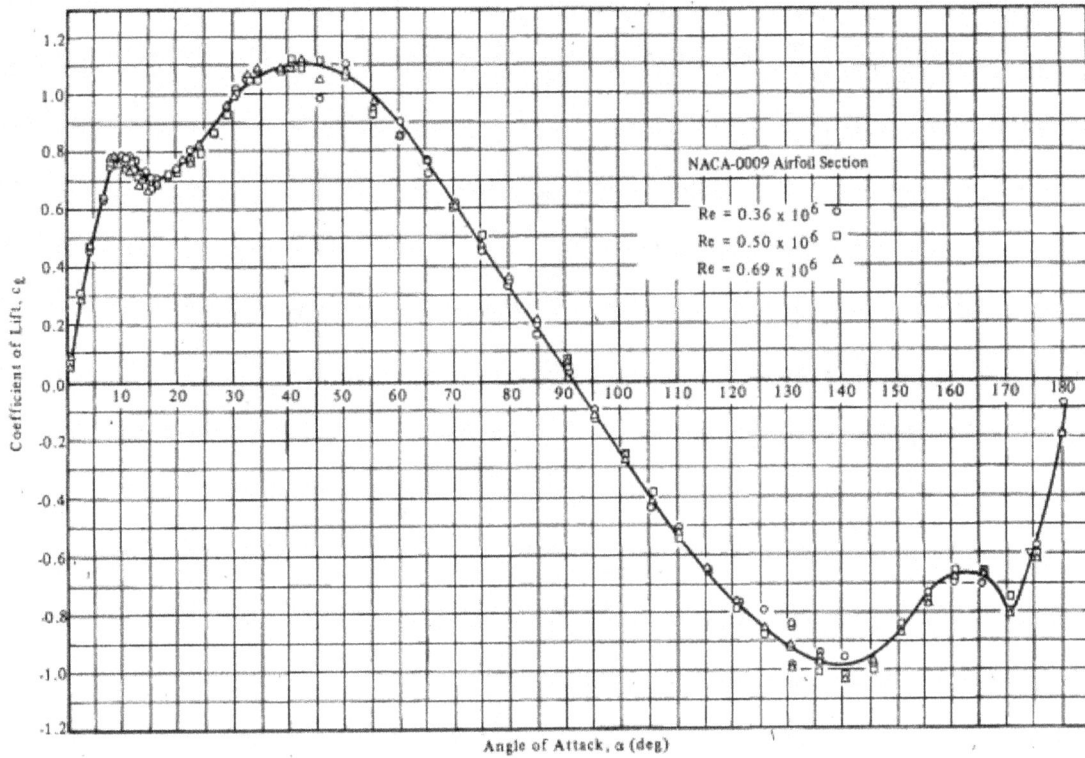

Figure. 7.2: *Variation of lift coefficient with angle of attack, of a symmetric airfoil from 0 to 180 degrees.From [93]*

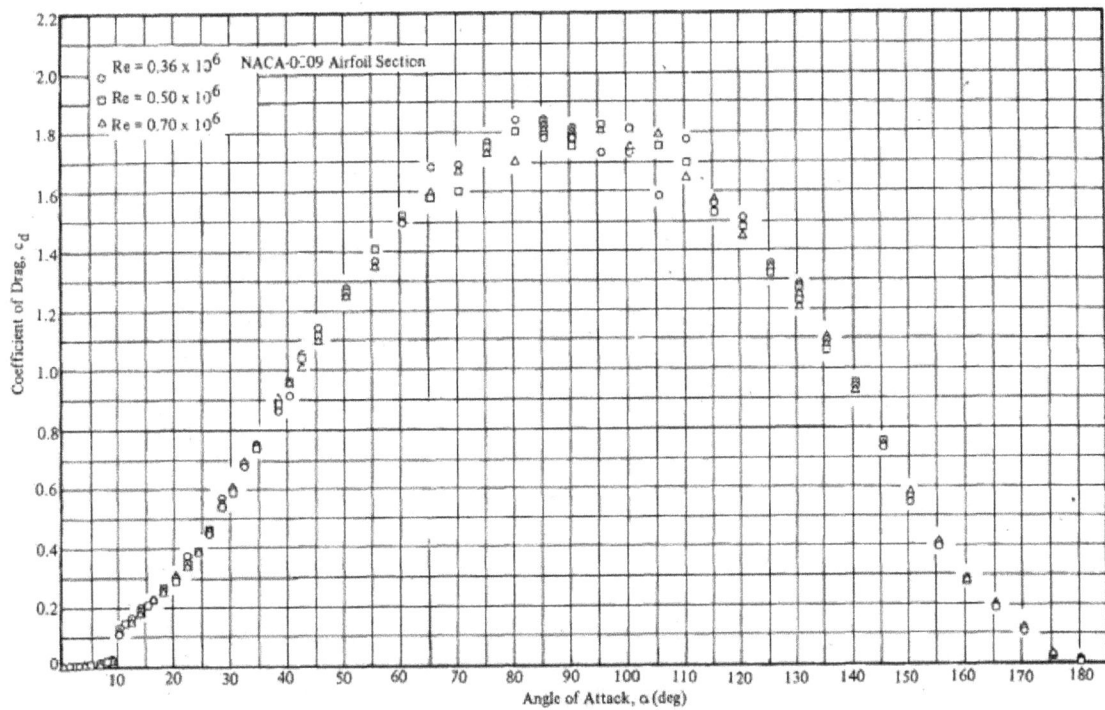

Figure. 7.3: *Variation of drag coefficient with angle of attack, of a symmetric airfoil from 0 to 180 degrees. From [93]*

approximately as square of angle of attack in the regime below 12 degrees, then again like a sine wave, reaching a maximum at 90 degrees angle of attack. However, compared to what happens beyond 12 degrees, the drag coefficient is very low between -12 and 12 degrees. Hence in this regime, the lift-to-drag ratio is high.

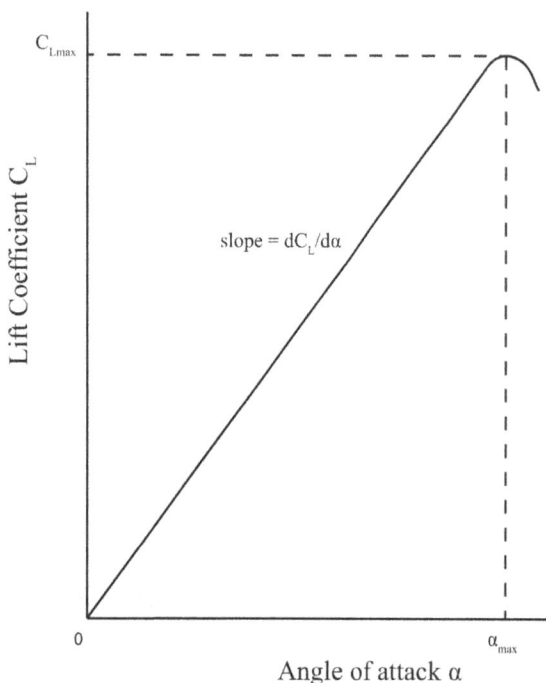

Figure. 7.4: The lift curve of a wing with symmetric airfoil cross-section.

The data in these two figures are from a Sandia Laboratories report [93] done in 1978, to help develop vertical axis wind turbines. The data were obtained at different values of Reynolds number. The Reynolds number is a measure of how unimportant the effects of viscosity are: the higher the Reynolds number, the less the effect of viscosity. Generally, larger Reynolds number means higher air speed, larger airfoil chord length, higher density (meaning low altitude in the atmosphere or colder day) and lower viscosity (lower altitude or colder day). The data shown are at Reynolds numbers of 360,000 to 700,000. There is not much effect seen.

If we look closely at these two variations, we see that the ratio of lift coefficient to drag coefficient reaches its highest value in the regime around 6 degrees. Thus in this linear regime of lift coefficient variation, we get the best lift to drag ratio. *This is the basis of most of aerodynamic flight: the possibility of obtaining high lift with low drag.* In the linear regime, the lift coefficient of a wing increases linearly with the angle of attack. The slope of this curve, $\frac{dC_L}{d\alpha}$, is called the lift curve slope. This is shown in Figure 7.4. Note that the lift coefficient does not rise indefinitely with angle of attack. After some α_{stall}, at which C_L reaches its max value, flow separation occurs along the wing. This is known as stall, and greatly reduces lift. In the C_L vs α graph shown in Figure 7.4 the lift coefficient is zero when the angle of attack is zero. We would say that the zero-lift angle of attack is 0 degrees. However, not all airfoils are symmetric. Some are cambered, which essentially means that they are not symmetric.

7.0.2   Vortex-Induced Lift

We have seen above that angle of attack and camber can generate lift. The third way to generate lift in low-speed flight is vortex-induced lift. The vortex generated at the wing tip is generally bad news, because it means lift loss and drag rise. However, being a vortex, it has regions of high velocity and low pressure. If we can make the vortex go close to the upper surface of the wing, this low pressure can provide the suction we need to generate lift. This principle is used on aircraft which, for other reasons, must have wings with extremely low aspect ratio. In fact

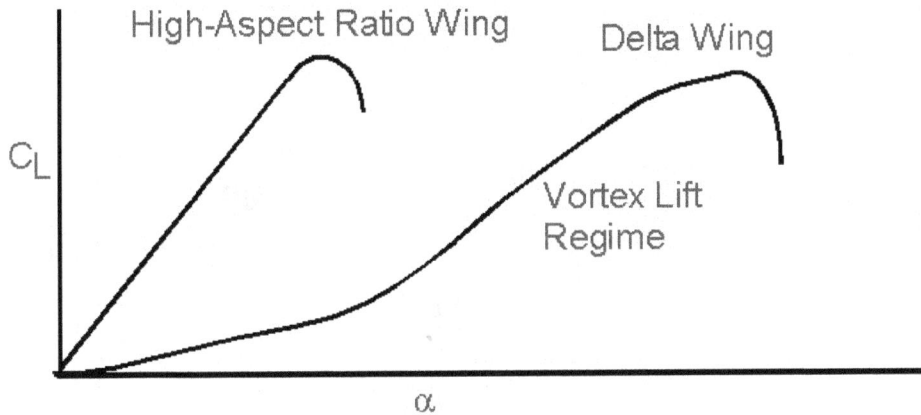

Figure. 7.5: Lift curve slope of a wing with vortex lift.

most aircraft designed for high-speed flight and high maneuverability have wings of small aspect ratio, with highly swept wing leading edges. The wing sweep is so high that we can think of the entire leading edge as the wing tip. Even at small angles of attack, a vortex forms along this edge (called, obviously, the Leading Edge Vortex), and this provides much of the lift of such wings when the aircraft is flying at low speed (even supersonic aircraft need to land, fairly slowly). When vortex lift is used, the wings can be very thin, and have sharp leading edges, which are good to minimize shocks and wave drag in high speed flight.

The lift-curve slope for cases of vortex lift is very small compared to the ideal lift curve slope of 2π per radian. However, vortex lift can be obtained up to large angles of attack, sometimes up to 30 degrees angle of attack. So adequate lift can be obtained by going to high angles of attack during landing and low-speed flight. The North American XB-70 supersonic bomber, the British Aerospace - Aerospatiale Concorde, and the Soviet Tupolev Tu-144 supersonic jetliners are examples of delta-winged aircraft. The delta wings are good for supersonic flight. When the aircraft comes in for a landing, it does so at a high angle of attack where the wings produce vortex lift.

7.1   Pressure

7.1.1   Bernoulli's Principle

In each lift generation method presented in the "Lift Generation" section, the flow moves more rapidly at some places than at others. In these regions of high velocity, the pressure is lower. The relation between pressure and velocity in low-speed flow is given by the Bernoulli equation [94–97]:

$$p_0 = p_1 + \frac{1}{2}\rho U_1^2 \tag{7.1}$$

$$p_0 = p_1 + \frac{1}{2}\rho U_1^2 = p_2 + \frac{1}{2}\rho U_2^2$$

Figure. 7.6: (Left)North American XB-70 landing with parachutes to slow down the vehicle after touchdown. (Right) Tupolev TU-144, used as a NASA testbed, shown at landing with nose drooped. Both images courtesy NASA

This equation is derived from Newton's Second Law of Motion, which expresses conservation of momentum. p_0 is called the stagnation pressure, or total pressure, while p is called the static pressure. The term $\frac{1}{2}\rho U^2$ is called the dynamic pressure, also denoted as q.

7.1.2   Pressure Coefficient

The pressure coefficient is a way to express the pressure with respect to some reference pressure, as a "dimensionless" quantity.

$$C_p = \frac{p - p_\infty}{\frac{1}{2}\rho U_\infty^2} = \frac{p - p_\infty}{q_\infty} = 1 - (\frac{U}{U_\infty})^2$$

$C_p = 0$ indicates the undisturbed freestream value of static pressure.
$C_p = 1$ indicates a stagnation point.
$C_p < 0$ indicates a suction region.
 Chordwise pressure distribution over an airfoil in low-speed flow.

Small as it looks, this figure is crucial to understanding aerodynamics. Look carefully at it. Firstly, it is plotted seemingly upside-down: the positive values of pressure coefficient go downwards, the negative ones go upwards. This is because in the days when people had to plot graphs by hand, they had to plan out their day's work carefully. They would take a fresh sheet of (expensive!) drawing paper and lay it very carefully aligned to be horizontal on their drawing table, using their T-square. Then they would measure out the margins to plan the dimensions of their plot. Things would be so much easier if the bottom boundary were carefully defined. And it is! The positive values of pressure coefficients in the equation above, would not exceed 1.0. The negative ones on the other hand could go pretty deeply negative. So they set up the convention that the positive pressures would go BELOW the horizontal axis. Besides, the term *Suction Peak* would be an oxymoron, would it not?

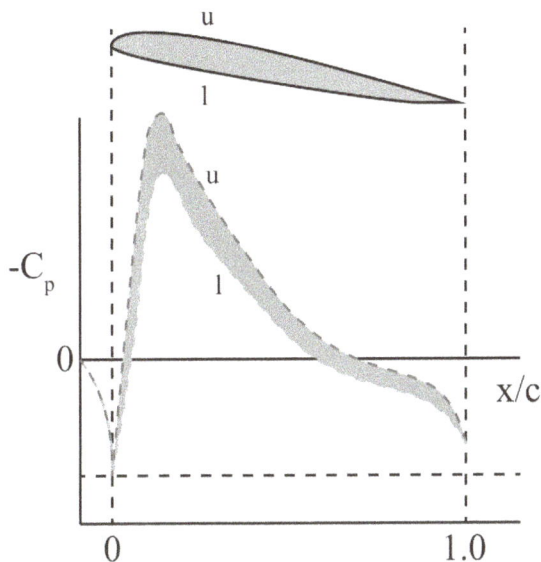

Figure. 7.7: Chordwise pressure distribution over an airfoil in low-speed flow.

Secondly, note that the upper surface and lower surface plots are SIMILAR, not mirror images! That's 20 points on a midterm test right there if you sleep through that. Why? Well, why not? If it were a symmetric airfoil at zero angle of attack, the flow along the upper and lower surfaces would see exactly the same changes in pressure at corresponding values of x/c. At a small angle of attack, there would be a small difference. Which is what we see. The entire lift comes from this DIFFERENCE. That is the shaded region shown in the figure. Note that much of the difference comes from the front region of the airfoil: the rear half contributes very little indeed. Why even have that half? OK, there's something for you to invent in your career: an airfoil that has only a front, and a sharp trailing edge, nothing in between.

Thirdly, note that the pressure decreases (magnitude of suction increases) until about the quarter-chord of the airfoil, then drops sharply. Both are significant. For thick, nicely rounded airfoils like the one shown, the suction peak rises sharply, and then drops sharply. This is not good news. The drop means the pressure is rising fast as the flow comes down the back side of the airfoil upper surface. To the flow, it does not feel like going downhill: it starts Feeling The Pressure, starts panting and sweating and says: " *Do I RE-ALLY want to keep going? How about climbing away from this surface and shooting off straight downstream? Or even curling back upstream?*" Perish the thought! This is Flow Separation. Recirculation. Stall!

Finally, we see that at the trailing edge, somewhat miraculously, the pressure coefficient rises back towards 1.0. Yes the trailing edge is a stagnation line. Ideal flow analysis would tell you that this means that the pressure there is the same as the upstream stagnation pressure, and for now, with the Bernoulli equation, we will accept that. However, in real life, there is friction at the surface. Some of the stagnation pressure is lost to heating and otherwise the chaos generated as the flow grates along the surface. So the stagnation pressure at the trailing edge is a bit lower than that at the leading edge. How much lower? Ha! a rather complicated answer that relates to the drag of the airfoil, whether there slow separation, turbulence, etc etc. Let's not go there. We can get airfoil data on airfoil drag and let us surf right over this question.

Exercise: Pressure Coefficient

What is the pressure coefficient at the stagnation point of an airfoil section?
What is the pressure coefficient on a flat surface aligned with the freestream?

C_p at the suction peak of an airfoil is -1.2. What is the pressure there as a percentage of the freestream static pressure?

What is the velocity at this point, as a percentage of the freestream velocity?

7.2   Airfoil (British: "aerofoil")

"Airfoil" means "shape of a section of a wing". It is a two-dimensional concept. Airfoils cannot fly; wings fly. Airfoil properties are used to calculate and design wing properties. The airfoil lift coefficient, also called the section lift coefficient, is denoted as c_l. Note that this lowercase coefficient is not the same as the uppercase version, C_L, which refers to the wing lift coefficient. As with the wing lift coefficient, the airfoil lift coefficient also varies with angle of attack α. If the airfoil is cambered, the lift coefficient is positive even at zero angle of attack, and reaches zero only at some negative value of α. This is called the "zero-lift angle of attack", α_0. As the camber increases, α_0 becomes more negative. Thus the airfoil coefficient is $c_l = \frac{dc_l}{d\alpha}(\alpha - \alpha_0)$. The lift curve slope is $\frac{dc_l}{d\alpha}$. $\frac{dc_l}{d\alpha} \leq 2\pi$, where α is in radians.

7.3   Example Problems

1. **Example: Airfoil Lift Coefficient**: The angle of attack of an airfoil is 12 degrees. The lift curve slope is 5.8 per radian. Zero-lift angle of attack is -2 degrees. Find the lift coefficient. *Answer: 0.4511*

2. If the air density is 0.9 of sea-level standard, and the temperature is 20 deg. C higher than the standard sea-level, flight speed is 100 m/s, and the airfoil chord is 1.2m, find the lift per unit span of the airfoil section of the previous problem at the conditions mentioned there. 7 *Answer: Dynamic Pressure is 5512.5 Pascals. Lift per unit span is 2984 N/m.*

3. **Example: Stagnation Pressure** A Pitot tube is used to measure the stagnation pressure in an air stream where the flow speed is 10m/s. The static conditions are sea-level standard: temperature of 288K, pressure of 101325 $\frac{N}{m^2}$. Find the stagnation pressure at the probe tip. Universal gas constant is 8313 in SI units. Molecular weight of air is 28.97. *Answer: Density should come out to be 1.225 kg/m^3. Dynamic pressure is 61.25 Pascals. Stagnation pressure is 101386 Pascals.*

4. **Example: Static Pressure** A Pitot tube is used to measure the stagnation pressure in an air stream where the flow speed is 100m/s. The density is 1.1 $\frac{kg}{m^3}$. At the stagnation point, pressure and temperature are 102,000 $\frac{N}{m^2}$ and 300K respectively. Find the static pressure. *Answer: Dynamic pressure q is 5500 Pascals. Static pressure is 96500 Pascals.*

5. **Example: Pressure Coefficient** At a point above the wing of an aircraft, a pressure sensor measures a static pressure of 50,000 $\frac{N}{m^2}$. The air temperature is 265K. The freestream speed is known to be 100 m/s. Find the pressure coefficient, if the freestream static pressure is 55, 000 $\frac{N}{m^2}$.

   Answer: q_∞ is 3287 Pascals. C_p is -1.521

6. **Example: Lift Per Unit Span** An airfoil of chord 0.5m has a zero-lift angle of attack of -2 degrees. The lift curve slope is 5.89 per radian. It is set at 5 degrees angle of attack in a flow of air, with density $1\frac{kg}{m^3}$, and speed 100m/s. Find the lift per unit span, and be sure to specify the units. *Answer: Lift coefficient is 0.7196. Lift per unit span is 1799 N/m.*

7. The straight wings of a civilian aircraft (no fancy fighter wings) has an airfoil section with the ideal lift curve slope of 2π per radian. If the zero-lift angle of attack is -3 degree, find the lift coefficient at 10 degrees angle of attack.

8. At a point above the surface of an aircraft, the pressure coefficient is -1.5. The aircraft is flying at 20,000 feet ISA, at Mach 0.25. Find the surface pressure at the point in question.

9. What is the stagnation pressure in the above problem?

10. An aircraft cabin is at 5000 feet pressure altitude, and the inside temperature is 70 deg. F. The aircraft is flying at 7,000 feet ISA, quite slowly. Suddenly a window blows out, and air starts flowing out. What is the stagnation pressure of this flow coming out? How fast will the flow become, so that its static pressure equals the outside static pressure?

11. A Pitot tube is used to measure the stagnation pressure in an air stream where the flow speed is 10m/s. The static conditions are sea-level standard: temperature of 288K, pressure of 101300 $\frac{N}{m^2}$. Find the stagnation pressure at the probe tip. Universal gas constant is 8313 in SI units. Molecular weight of air is 28.9.

12. At a point above the wing of an aircraft, a pressure sensor measures a static pressure of 50,000 $\frac{N}{m^2}$. The air temperature is 265K. The flow speed is known to be 100 m/s. Find the air density and the stagnation pressure.

Chapter 8

More on Aerodynamics

8.1   Introduction

Unlike airfoils, studied in the last chapter, wings are 3-dimensional objects. They have a span, b, which is simply the distance from one wingtip to the other. They also have an aspect ratio, AR, defined by:

$$AR = \frac{b^2}{S}$$

where b is the span and S is the wing planform area. It is important to note that airfoils are considered to have an infinite span, and an aspect ratio of infinite.

8.1.1   Effects of Finite Aspect Ratio

At the ends of the wings, the pressure difference between the upper and lower sides is lost, as the flow rolls up into a vortex. This does not happen with airfoils, because, as previously stated, their spans are considered infinite and thus the flow never rolls up.

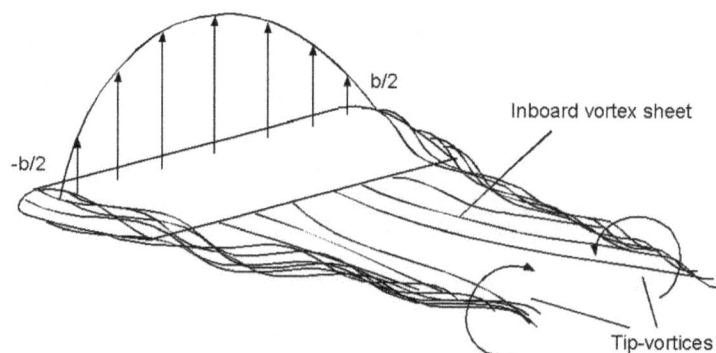

Figure. 8.1: Wingtip vortices.

Wing tip vortices were beautifully described by Lanchester [37, 98–100]as being similar to ropes twirling around each other. More to the point, like tendons of muscles wrapping around each other. He thus described vortices as the *sinews of fluid motion*. This lent support to the idea of calculating the rest of the flowfield properties as being *induced* by vortices. The effect of this vortex system is to cause a downward (or opposite to the general direction of lift) flow at the wing. This is viewed as being associated with the decrease in overall lift relative to the airfoil lift value predicted for a section of an infinite wing. The lift vector is tilted back, so that an induced drag is created. This is a pretty neat ('slick' might be a better term) way of explaining something that is far more complex in origin. Why is lift actually lost? It is because the pressure difference between the upper and lower surface flows, decreases outboard and disappears at the wing tip. Why is drag experienced? Because some of that energy that you put into driving the wing forward, is now tied up in making the flow spin round and round - contributing nothing to forward motion. But both of these ideas become hopelessly complicated to use for calculations. The 'Look the Effective Lift Vector Is Tilted Backwards!' interpretation was an immense breakthrough towards calculating the lift and drag of a wing. Instead of the vortex system being induced by the wing, the flow around the wing is now considered to be induced by the vortex system.

The actual theoretical formulation along these lines of 'induced velocity' is attributed to the brilliance of Professor Ludwig Prandtl [101–103]. Of course this was some 20 years or more before Lanchester made his beautiful drawings. In fact he managed to reduce the entire lifting effect of a long (high aspect ratio) wing to the effect of a line vortex of varying strength, placed approximately along the quarter-chord line of the airfoil sections of the wing. More on this in your aerodynamics courses.

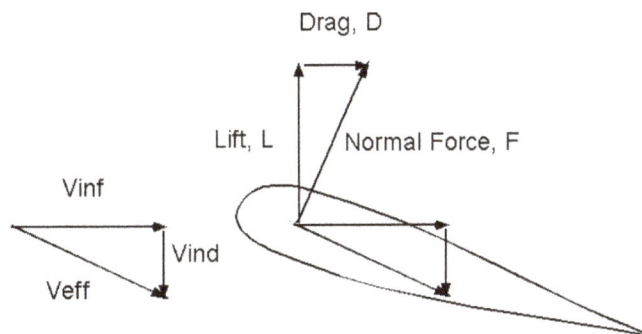

Figure. 8.2: Lift loss and induced drag can be explained simply by considering an effective freestream velocity that is the vector sum of the actual freestream velocity and the velocity induced by the trailing vortex system.

Both of these (usually undesirable) effects are reduced by increasing the aspect ratio of the wing. As the aspect ratio approaches infinity, the wing resembles an airfoil, and will be less affected by these effects.

8.2   Lift Coefficient

Due to the effects of finite aspect ratio described above, the lift curve slope of a wing, $\frac{dC_L}{d\alpha}$, will be smaller than the lift curve slope of the airfoil shape it uses. If we set the airfoil's lift curve slope to $a_0 = \frac{dc_l}{d\alpha}$, we can find the lift curve slope of a wing using this airfoil with the following equation:

$$\frac{dC_L}{d\alpha} = \frac{a_0}{1 + \frac{a_0}{\pi ARe}}$$

The lift curve slope of a wing depends on the lift curve slope of its airfoil as well as its aspect ratio and spanwise efficiency factor. As the equation above shows, any wing with a finite aspect ratio will always have a smaller lift curve slope than its airfoil. As the aspect ratio of the wing approaches infinite, the lift curve slope of the wing approaches that of the airfoil. An aspect ratio of infinite would yield

$$\frac{dC_L}{d\alpha} = \frac{a_0}{1 + 0} = \frac{dc_l}{d\alpha}$$

In this chapter's introduction, we stated that airfoils are considered to have an infinite aspect ratio. The results of this equation agree with this.

8.3   Drag Coefficient

Drag is given by

$$D = \frac{1}{2}\rho U_\infty^2 S C_D$$

The drag coefficient in low-speed flow is composed of three parts:

$$C_D = C_{D0} + C_{Dfriction} + C_{Di}$$

where C_{D0} is the parasite drag coefficient, $C_{Dfriction}$ is the skin friction drag coefficient, and C_{Di} is the lift-induced drag. These three terms make up what the total drag coefficient, C_D, for low-speed flight.

8.3.1   Parasite/Profile Drag

The term C_{D0} in the drag coefficient equation above is the parasitic drag coefficient, also known as the profile drag coefficient. This term is independent of lift. It is usually due to the losses of stagnation pressure which occur when part of the flow separates somewhere along the wing or body surface. In high speed flight, the effects of shock and wave drag must be added to this, and becomes the dominant source of drag. Most aircraft are designed to minimize C_{D0}.

The profile drag of an airfoil of chord 1 unit is about the same as that of a circular cylinder whose diameter is only 0.005 units.

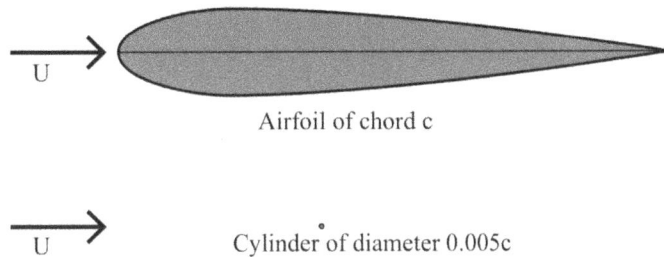

Airfoil of chord c

Cylinder of diameter 0.005c

Figure. 8.3: The profile drag of an airfoil is about the same as that of a circular cylinder whose diameter is 0.5% of the chord length.

8.3.2   Skin Friction Drag

$C_{Dfriction}$ is the skin friction drag coefficient, which is due to viscosity. This becomes important in two limits: one where the size of the wing, or the speed of the flow, is extremely small, as might be the case for an insect-sized aircraft. This is called the "low-Reynolds number" limit. The other limit is that of high-speed flight, where the skin friction can be severe enough to heat up the wing surface to melting point.

8.3.3   Induced Drag

C_{Di} is the lift-induced drag coefficient. In low-speed flight, this is the largest cause of drag, because you must have lift to fly, and this drag is caused by lift. The lift induced drag coefficient can be calculated with the following formula:

$$C_{Di} = \frac{C_L^2}{\pi A R e}$$

In the above equation, AR is the aspect ratio, C_L is the lift coefficient, and e is the spanwise efficiency factor. The spanwise efficiency factor answers the question: How does this wing rate compared to the ideal wing for this aspect ratio? Its value is usually close to 1, perhaps as high as 0.99. If unknown, it can usually be assumed to be anything from 0.9 to 1.

Two things should be noted from the equation for the induced drag coefficient:

1. $C_{Di} \propto C_L^2$ so that $C_{Di} \propto \alpha^2$. Induced drag increases with the square of the angle of attack. A smaller angle of attack reduces the induced drag.

2. C_{Di} approaches zero as aspect ratio approaches infinite. A larger aspect ratio reduces the induced drag.

To minimize induced drag, one should design wings with the largest possible aspect ratio, but also provide enough surface area so that you only need a small angle of attack to provide the necessary lift even at low speed.

Example: Calculating Drag

C_{D0} of a small airliner is 0.018. The wing aspect ratio is 6. Assume spanwise efficiency is 0.9. The lift coefficient is 0.5. Find the total drag coefficient. If the density is 1 kg/m^3, the span is 40 meters and the speed is 200 m/s, find the drag.

8.3.4   Speed for Minimum Drag

As mentioned before, total drag is composed of a part that depends on lift, and one that does not.
$$D = D_0 + D_i = (C_{D0} + C_{Di})(\frac{1}{2}\rho U_\infty^2 S)$$
$$D = (C_{D0} + \frac{C_L^2}{\pi(AR)e})(\frac{1}{2}\rho U_\infty^2 S)$$

Let us consider what it takes to keep lift equal to drag, L = W: $W = L = q_\infty S C_L$ where $C_L = \frac{W}{q_\infty S}$ So
$$D = q_\infty S C_{D0} + (\frac{W}{S})^2 \frac{1}{\pi(AR)e}(\frac{S}{q_\infty})$$

To find the minimum of a function, one should differentiate the function and set the derivative equal to zero. Strictly speaking, the Calculus tells us that we are finding an *extremum* and we should differentiate again, find the second derivative and check its sign and see if it is positive or negative to determine whether the extremum is a minimum or a maximum. OK, we are pretty sure that the speed for MAXIMUM drag is infinity, so we will comfortably assume that the extremum that we find is in fact a minimum. (Hope it does not come out to be 10,352.85 km/s as my dear student wrote in a test.)

$$\frac{dD}{dq_\infty} = S C_{D0} - (\frac{W}{S})^2 \frac{1}{\pi(AR)e}(\frac{S}{q_\infty^2}) = 0$$

$$C_{D0} = (\frac{W}{Sq_\infty})^2 \frac{1}{\pi(AR)e}$$

$$C_{D0} = C_{Di}$$

Minimum Total Drag = twice zero-lift drag. This is a remarkable result. It means that aircraft, unlike other forms of transportation, have a definite speed for minimum drag.

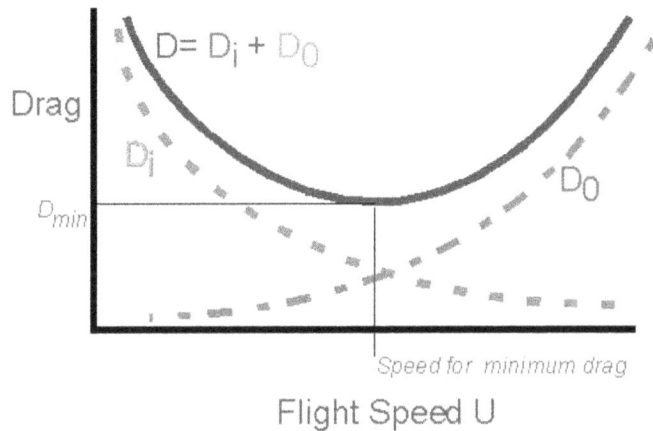

Figure. 8.4: For aircraft, the speed for minimum drag is nonzero.

To fly an airplane of a given weight, straight and level, the condition for minimum drag (maximum lift-to-drag ratio) is that the profile drag coefficient is the same as the induced drag coefficient.

Example

An aircraft has a wing loading (W/S) of 130 pounds per square foot (6233 N/m^2), aspect ratio of 7.667, and wing span of 60.96 m. We' will assume that its spanwise efficiency factor will be 0.99. Let us assume that the profile drag coefficient is given by $C_{D0} = 0.015$.

Thus, for maximum Lift-to-Drag ratio (minimum drag, and lift is always equal to weight for straight and level flight),

$$C_{Di} = C_{D0} = 0.015$$

The corresponding C_L is calculated as 0.598, and the dynamic pressure is 10423 N/m^2.

At 11,000 meters in the Standard Atmosphere, density is 0.36 kg/m^3, so that the flight speed is 240.64 m/s.

Note: In practice, the C_{D0} might change with flight Mach number, for high-speed flight. This is not taken into account in the above example.

8.4   Aerodynamics Summary

Lift is the force perpendicular to the flow direction, due to pressure differences across surfaces. There are 3 ways of generating lift:

1. angle of attack

2. camber

3. vortex-induced lift

An infinite-span (2-dimensional) wing is entirely described by its airfoil section.

Finite wings have less lift than corresponding infinite wings at the same angle of attack, and also have lift-induced drag. The total drag acting on a wing in inviscid flow is composed of profile drag, which does not vary with lift, and induced drag, which rises with the square of the lift coefficient. The friction drag is ignored in the inviscid case.

To fly an airplane of a given weight, straight and level, the condition for minimum drag (maximum lift-to-drag ratio) is that the profile drag coefficient is the same as the induced drag coefficient.

8.5   Going Further

What we have discussed above is a very 'bare-bones' introduction to aerodynamics, ignoring everything except the regime of low-speed flight at moderate Reynolds numbers typical of small aircraft. The vast ocean of high-speed flight, and the growing flood of work on very low Reynolds number, as well as the largely unexplored field of bluff-body aerodynamics (everything that is not nice and streamlined and pointy-tailed) all remain out there, but what is given above should still stand you in good stead as you wade into those. Hundreds, perhaps thousands, of books have been written on Aerodynamics. The number of research papers is probably high enough to build a ladder reaching the top of the atmosphere, and getting higher very fast. Let's start with a very few [7, 10, 29–32, 34, 35, 37–39, 104–107]. To learn about airfoils, one need go no further than Abbott and von Doenhoff [30], an extensive work compiled in the 1950s. More recent work of course exists, dealing with supercritical airfoils [108–112] for high-speed flight, and low-Reynolds number airfoils [113–115] for sailplanes [116–118], gliders [119–121] and small vehicles [122].

What follows is a brief Concept Module that was developed to assist learners in the Aerospace Digital Library under Project EXTROVERT, the same NASA-funded project under which the first version of this book was written, as an e-book collection of slides from the course. It is reproduced here by permission of the author (i.e., myself).

8.5.1 Definition

When objects move through air, forces are generated by the relative motion between the air and the surfaces of the object. Aerodynamics is the study of these forces, generated by the motion of air.

8.5.2 Introduction

Aerodynamics deals with the forces due to the motion of air and other gaseous fluids relative to bodies. Aerodynamic lift is generated perpendicular to the direction of the free stream as the reaction to the rate of change of momentum of air turning around an object, and, at high speeds, to compression of air by the object. Flow turning is accomplished by changing the angle of attack of the surface, or by using the camber of the surface in subsonic flight, or by generating vortices along the leading edges of swept wings. In flight at high speeds, the pressure changes generated when the oncoming flow is slowed down or accelerated around the object, causes large changes in density between different sides of the object, again resulting in a net change in momentum. The forces generated as a reaction to this rate of change of momentum, cause lift in high speed flight.

All lift generation is accompanied by some generation of drag forces opposing the motion of the object through air. Various mechanisms of drag generation have been identified. Achieving the highest possible ratio of lift to drag is a usual goal of aerodynamic design, except of course in cases such as parachutes and speed brakes where high drag is desired.

The behavior of air in motion can be described in general terms using physical theories at various levels, going from the dynamics of huge masses of air such as hurricanes, down to the tiniest scales of atomic motion. However it is unnecessary to use these general, all-inclusive theoretical descriptions to solve most problems. To design vehicles and predict their performance, we use several methods, each of which is restricted to a small range of parameters. Thus, for example, we divide the field of aerodynamics into categories based on the speed range of interest. The behavior of air flows changes depending on the ratio of the flow speed to the speed of sound. This ratio is called the Mach number. The speed of sound is the speed at which information propagates through a gas. So if the vehicle moves faster than the speed of sound, the air ahead of it cannot "move away": there is no way for it to "know" of the approaching vehicle. This leads to the formation of "shock" waves in the air ahead of the vehicle. Very close to a body surface, or at the interface between two streams of air moving at different speeds, we encounter friction. This leads to many strange and beautiful effects, producing the sinuous structures which make us want to keep looking at flowing streams for hours. Unfortunately, these things are quite difficult to calculate, so we argue that the primary effects of friction are confined to a region very close to the surface, called the "boundary layer". The boundary layer is a "shear layer". Likewise, the region between two streams of air, flowing at different speeds, is called a "free shear layer" because no solid surface boundary is involved. Away from surfaces, the flow can usually be considered to the "inviscid": it is almost as if viscosity does not exist there.

8.6   Classification of speed ranges by relation to the speed of sound

Regimes of Mach number are classified as:

1. Incompressible; Low Speed Aerodynamics ($0 < M < 0.33$)

2. Subsonic ($0.33 < M <$ Mcritical)

3. Transonic Aerodynamics (Mcritical $< M < 1.2$)

4. Supersonic Aerodynamics ($1.2 < M < 4$)

5. Hypersonic Aerodynamics ($4 < M < ?$)

6. Relativistic Aerodynamics, where the speed becomes comparable to the speed of light.

8.6.1   Low Speed Aerodynamics ($0 < M < 0.33$)

Here the speed range is from zero to roughly 1/3 the speed of sound. The speed of sound in the atmosphere is roughly 340 meters/second, so low-speed aerodynamics covers speeds of 0 to roughly 100 m/s. What is special about this range? The density of the air (mass per unit volume) does not change appreciably due to changes in velocity of this magnitude (i.e., from 0 to 0.3 times the speed of sound). The maximum variation in density is less than 5% of the value of density. Thus, in this speed range, the flow is said to be "incompressible" (by changes in velocity!). With this assumption, we can treat air flow in a manner similar to water flow over a body. One primary simplification resulting from neglecting changes in density, is that the velocity field around an object can be completely described by considering just the fact that mass must be conserved. This can be done whether the flow problem is steady or time-dependent. Once this continuity equation is solved for the velocity field, the pressure at every point can be calculated directly using conservation of momentum. From the pressure distribution around the object, the net force can be resolved into lift, drag and side force, and the moments about the different axes can also be calculated.

8.6.2   Subsonic Aerodynamics ($0.33 < M < $ Mcritical)

Here the speed range is from about 1/3 the speed of sound, to about 0.8 times the speed of sound. When vehicles move in this speed range, the flow variations occurring over the vehicle surfaces involve substantial density variations. This effect must be taken into account in performing calculations, or the results obtained will be quite wrong. The upper limit of this regime is the flight Mach number where the local flow somewhere over the aircraft becomes sonic. This flight Mach number is called the "critical Mach number". It depends on the aircraft configuration, and the attitude at which it is flying. Flying faster than the critical Mach number makes the flow supersonic over some part of the aircraft. When this flow decelerates, shocks are produced, with a large increase in drag.

In the subsonic regime, calculating the pressure distribution requires solving for the velocity, pressure and density simultaneously, from the equations describing conservation of mass, momentum and energy. A much simpler practical approach for the small flow perturbations typical of good aerodynamic designs, is to find transformations of coordinate systems, so that a *corresponding incompressible flow problem is solved in the new coordinate system*, and the solution transformed back to the coordinates of the subsonic flow problem. This is made possible by identifying the dependence of all properties on the flight Mach number. The Prandtl-Glauert transformation allows engineers to relate pressure and lift coefficients at a subsonic Mach number for a given configuration, to the corresponding values under incompressible flow conditions.

8.6.3   Transonic Aerodynamics (Mcritical $< M < 1.2$)

Most of today's airliners fly at speeds very close to the speed of sound. Today's engines work very well in this regime, and today's people want to reach their destinations quickly and as cheaply as possible. However, this is a very difficult flow regime to analyze, because the changes occurring over an aircraft flying at transonic speeds involve changes from "supersonic" to "subsonic" and back.

8.6.4   Supersonic Aerodynamics ($1.2 < M < 4$)

The behavior of flows moving faster than the speed of sound is very different from that of flows moving slower than sound. The simple explanation for this is that sound cannot propagate upstream in such flows; so these flows cannot "know" of changes about to occur further downstream. Changes occur very suddenly, and through distinct flow features, rather than the curves and gradual changes of subsonic flows. The solutions to the conservation equations then take the form of waves, propagating at the speed of sound.

8.6.5   Hypersonic Aerodynamics ($4 < M < ?$)

Nothing magical happens at Mach 4, but given the decisions that designers make, Mach 4 is approximately the regime where hypersonic effects become a significant concern. As the Mach number increases, the changes caused by deceleration become very large. When a high-Mach number flow is stopped, say at the nose of a vehicle, the temperature, pressure and density increase by large amounts. This increase may be large enough that the properties of the air, such as the specific heat and even the molecular structure, change. This is generally considered to become significant above Mach 4. In the hypersonic regime, the disturbance caused by the aircraft is not felt until the vehicle is very close indeed: the "shocks" lie so close to the surface that the layer of air between the shock and the vehicle is quite thin. The concept of Mach number begins to lose significance as these changes occur, and engineers resort to descriptions in terms of "enthalpy" rather than Mach number to deal with flows at very high speeds. We can still take the flight speed and divide by the speed of sound in the undisturbed atmosphere, and arrive at a flight Mach number. For spacecraft re-entering the atmosphere without aerodynamic controls (such as the Apollo capsules) this "Mach number" was about 35; the Space Shuttle glides in at around Mach 25; meteors might come in at Mach number of 250 or more.

8.6.6   Relativistic Aerodynamics

No human-built object has started flying in this range; however, it is easy to think about a star which is part of one galaxy encountering another galaxy moving at a very different speed. Here the relative speeds may become a significant fraction of the speed of light. The flows inside the engines of spacecraft may reach such speeds, as engineers explore propulsion devices to power spacecraft towards other stars.

8.6.7   Classification of Flows according to properties of significance

Engineers often use terms which are most relevant to the methods which they are using to describe a particular problem and solve it. The following list is certainly not all-inclusive, but illustrates the reasoning behind the terms. The terms are not mutually exclusive. Thus, one speaks of a steady turbulent reacting flow, or an unsteady, high-temperature potential flow. One cannot speak of a laminar turbulent flow, or a steady unsteady flow, obviously.

Steady Flows: This term is used to describe situations where there is no rapid change in properties over time. For example, an aircraft cruising straight and level in the upper atmosphere, well above where any gust can reach it. When such situations are being analyzed, a lot of effort can be saved by neglecting terms in the equations which describe rates of change of the flow properties or forces at any point on the aircraft.

Unsteady Flows: Obviously, unsteady means "not steady", but it also means a lot more. Many situations of "not steady" can be made "steady" by appropriately changing coordinates. For example, an aircraft flying around in circles can be considered to be flying a steady, "coordinated turn", with the flow properties not changing at any point on the craft. The rotor blade of a helicopter hovering steady in still air also sees the same flow properties from instant to instant. Now if the aircraft starts rising or sinking, or stays in the coordinated turn long enough to burn a lot of fuel and thus require changes to the control surfaces and engine thrust, or if the helicopter starts flying forward very slowly, there is a small rate of change of properties encountered, but this is slow enough to be broken up into several stages of steady flow. Likewise, if a wing flaps up and down very slowly, its attitude and the forces on it, change with time, but this can be analyzed by breaking the flapping motion into several steady steps and "joining the dots" of the answers at each step. On the other hand, an aircraft wing encountering a gust is a situation which requires "unsteady aerodynamics". A helicopter rotor at moderate forward flight speed encounters substantially different flow conditions within each revolution. A shock forming at the inlet of a jet aircraft, and then disappearing, causes large unsteady effects. The flapping motion of an insect's wings is an extreme case of unsteady aerodynamics. In all these cases, the fact that there is a high rate of change has an important bearing on the results.

Inviscid Flows: To describe flows which are away from surfaces, one can simplify the theory and neglect the influence of viscosity of the fluid. Such descriptions are adequate for situations where the flow velocity and the size dimensions are large. The resulting methods are adequate to explain most of the lift forces (forces perpendicular to the freestream direction) on vehicles; they can also explain the formation of parts of the drag forces, but not all.

Potential Flows: When viscosity is neglected, and the effects of any rotation and shear in the flow are replaced by mathematical artifacts known as "singularities", the behavior of the rest of the flow can be analyzed by methods similar to those used to analyze electric fields and magnetic fields. These methods of "potential theory" are very powerful: they are used to do the initial calculation of the air loads on the wings, rotors and fuselages of most airplanes flying today, and also to analyze what happens when the flow is unsteady.

Viscous Flows: When flow close to a solid surface, or near any boundary where there is relative motion, is analyzed, the effects of fluid viscosity become important. So, analyses of such regions of flows must use equations which include the terms describing the effects of viscosity.

Laminar Flows: Generally used in analyses of "viscous flow" problems, this term means that the flow is "smooth", and resembles layers of fluid with slightly different velocities. In such flows, the effects of forces due to viscosity are significant, when compared to the effects of the inertia of the fluid motion. In other words, the "Reynolds number" which describes the ratio of inertial forces to viscous forces, is not very large (it still can be of the order of 100,000, but probably not 1 million.) In describing such flows, it is possible to arrive at an answer to the question: "what will the velocity be at this point one second from now?" to a high degree of accuracy: unlike "turbulent flows", described below.

Turbulent Flows: When there is some source of shear between different regions of fluid, and the Reynolds number is extremely high (on the order of a million or more), flows become turbulent, with the velocity, flow direction, and all associated properties fluctuating from instant to instant. Analyses of such flows must thus account for the results of such fluctuations, which include increased skin friction drag, reduced occurrence of flow separation and its drag, different rates of heat transfer between the flow and vehicle surfaces, the generation of noise, faster mixing between different fluid streams, and faster propagation of flames through gases.

High Temperature Flow: Over a narrow range of temperature, the properties of air, such as its specific heat, molecular composition etc. can be assumed to hold constant. We generally don't even worry about this issue in most of aerodynamics. However, there are situations where the changes in temperature are large, and hence we have to include detailed models of how gas properties change with temperature, in solving such problems. An example is the shock wave in front of the nose of the space shuttle as it comes down through the atmosphere. If we ignored the changes in gas properties in analyzing the change of temperature through this shock, we would get ridiculous results: we would predict temperatures which do not occur except in nuclear explosions!! As seen above, it is not the actual "highness" of the temperature that matters for this definition: it is the fact that the temperature can change over a large range.

Reacting flow: If the chemical reactions occurring in the flow are significant to the changes in flow properties, one has to include them in the analysis. For example, a "flame" is a reacting flow, where there is a large and rapid release of heat occurring during a chemical reaction. The molecular composition also changes, of course, during the chemical reaction. The flow in the combustion chamber of a rocket is a reacting flow. However, the flow in the nozzle of the same rocket, though it is glowing hot, may or may not be classified as a reacting flow: this depends on whether the composition changes substantially in the nozzle.

Non-equilibrium flow: This is another of those "when did we assume that?" revelations. In most problems in aerodynamics, we assume that we have "equilibrium" in the flow. The rates of collisions between molecules are high enough that we can assume, for example, that the temperature and pressure in the flow in a nozzle adjust instantly to changes in the nozzle geometry. In some flow situations, the changes in properties may be so large and so sudden that the flow has moved a significant distance before there is complete adjustment of the temperature and the chemical composition. This occurs in lasers, for example, where the medium is kept out of equilibrium. Non-equilibrium phenomena can cause important differences to the pressure distribution and hence the pitching moment on a high-speed aircraft, like the Space Shuttle at re-entry. Calculations of nozzle geometry and heat transfer for rocket engines and vehicles such as the X-33 also require non-equilibrium considerations.

Multiphase flow

Sometimes, flows may include changes between solid, liquid and gas states, and may also include substantial amounts of material of different phases flowing along together. For example, as compressed air at room temperature is expanded through the throat of a supersonic wind tunnel, some of the constituent gases may begin to liquefy, and droplets of oxygen or nitrogen might form in the flow. The flow over the leading edge of a rotor blade operating under icing conditions may involve the formation of ice particles near the surface.

Rarefied flow

In the upper reaches of the atmosphere, the density of the air becomes so low that air cannot be assumed to be a continuous medium or "continuum". The flow analysis must include consideration of this fact. Since collisions between molecules become rare, it is smarter to regard the flow as being composed of many balls bouncing off the surface of the vehicle.

When the gases in the flow are ionized (electrons leave many atoms, so that there is a high concentration of positive ions and free electrons), the flow behavior can be modified, and forces generated, using magnetic fields. Such flows are called plasmas.

Buoyant flow

These are probably the most common flows of air in the atmosphere: flows driven by the changes in their density due to heating or cooling, making them lighter or heavier than surrounding fluid. In most aerodynamics problems, we can neglect buoyancy effects because the flow velocities and the inertial effects are so large; however, buoyancy is a driver of atmospheric flows. Gliders, obviously, take advantage of buoyant flows when they rise on "thermals".

Supersets and Subsets of Aerodynamics Fluid dynamics and Aeromechanics and supersets of Aerodynamics. Hydrodynamics is a subset: this is mostly incompressible flow aerodynamics, with effects of buoyancy becoming important since it deals mostly with forces developed in liquids which are about a thousand times more dense than air. The speed of sound in liquids is very high, so that flows with speed comparable to the speed of sound are very rarely encountered. Gas dynamics is another subsets, which deals primarily with phenomena occurring over a wide range of Mach number. Other subsets are listed under the classifications given above.

8.6.8   Calculators/Applets

Hyperlinks are not given, because today Internet Search Engines and people have become quite adept and very quick at locating what they want, given some idea of what they should be seeking.

1. Standard atmosphere calculators

2. Shock calculators

3. Prandtl Meyer expansion calculators

4. Conical Flow calculators based on the Taylor-McColl scheme

5. The Wolfram Computational Document Format (CDF) site, where there are several example projects.

8.7   Analytical Codes

1. The XFOIL code developed by Professor Drela at MIT [115] calculates 2D airfoil aerodynamics at low speed, including the effects of boundary layers. This works well in the linear

regime of lift curve slope, but caution must be exercised in interpreting results beyond stall (as someone once remarked at an AIAA meeting, we try to keep our airplanes from stalling, so this is no criticism at all).

2. The Eppler code is extensively used in sailplane design (as is the XFOIL code)

3. The Murman-Cole Transonic Small Disturbance airfoil aerodynamics code is efficient enough to run on personal computers. There are many subsequent versions from other researchers that solve various issues.

4. Full Potential equation solvers are also efficient enough to be run on personal computers, and can deal with many aspects of subsonic and transonic flows, but have serious problems once strong shocks occur, or when flow separation occurs.

5. Euler solvers are based on a large finite difference grid, and hence the resources required grow explosively as the grid size increases.

6. Codes based on the Moretti scheme of solving the unsteady conservation equations to obtain steady solutions for blunt-body flows.

7. Navier-Stokes codes. These typically require high-speed computers such as the National Supercomputer facilities at various centers, but are becoming viable for increasingly realistic problems even on personal computers, particularly those equipped with fast Graphical Processor Units (GPUs). In other words, an excellent excuse to get a computer with the ultimate video game capabilities!

8.8   Design Update

OK, this is the point in the course where people start feeling sorry for themselves. So many concepts! So many EQUATIONS! Let us pause and update our design process and knowledge. What did we do so far?

1. We started with a Requirements Definition. How many passengers? How much cargo? What range? What speed is acceptable? After extensive Customer Discovery, we establish what might sell.

2. Next comes the calculation of Takeoff Weight (TOW). We started by calculating the payload with standard common sense values for the average weight per passenger, the number of crew needed, the weight of food and water to be loaded for a flight of so many hours.

We then used Benchmarking: We sought out what had been done in the past, to see the relation between the payload and range and speed and the ultimate Takeoff Weight. From this we estimated the highest Payload Fraction that could be achieved with all our smart advances and the experience gained from the passage of time. Using this Payload Fraction and the payload, we calculated our first guess of the TOW. We have rough estimates from benchmarking, of the weight fractions needed for the engines, and we know that no one can built a successful airplane with a structure that weighs less than, say, 25 percent of the takeoff weight if it is a subsonic aircraft, and 28 percent if supersonic. So we know that the maximum fuel weight fraction is what is left after all these are added up. It is a safe guess that if that fraction gets over 40 percent we are on thin ice. If it gets over 50 percent, we'll have to cut back on payload fraction. Just hard reality. The Concorde, with 55% fuel fraction, never made a profit despite extremely high ticket prices. The Soviet Tupolev Tu-144 pushed the limits a bit too much - and paid the price in fatal crashes and loss of confidence.

3. Next, we took a guess (again from benchmarking. What should be the Wing Loading (TOW divided by the Wing Planform Area S) for our class of aircraft? The numbers are fairly well established, but vary widely with the speed and altitude. We pick a reasonable value. That immediately allows us to calculate the wing planform area S. The wing span is limited by practical considerations, like being able to squeeze in between two Gates at the airport. Or maybe in the narrow confines of an aircraft carrier hold, even with the wings folded. The Aspect Ratio is defined as the square of span, divided by S.

4. We took a necessary detour into Atmospheric Science. We learned how to relate the properties of air to one another, and about the Standard Atmosphere model of how conditions change with altitude. We even figured out how to calculate the atmospheres of Mars and Venus. All in all, we found that calculating atmospheric properties at any altitude in Earth's Standard Atmosphere is ridiculously easy, and there is no need for tables and charts. Of course we also peeked at where there are convenient Calculators on the Internet that we can use with confidence.

5. Now you are into aerodynamics. Knowing the wing loading W/S, and the chosen altitude and speed, we can calculate the lift coefficient needed in cruise. A high lift coefficient for cruise, is bad news, because the Lift-Induced Drag rises as the square of the lift coefficient. So now we can calculate the lift and induced drag coefficients. There is also the part of the drag that does not depend on lift: this is the part that is due to flow separation and vortices generated at junctures (wing/fuselage, wing/engines etc) and the drag due to skin friction. Estimating the drag accurately is both necessary and difficult. We will do as good a job as our knowledge allows, but keep pushing on that knowledge to improve the sophistication of this calculation. This is one area where sophistication and accuracy ARE important. As the Boeing and Airbus people might say, uncertainty by one digit in the fourth decimal place of the drag coefficient, is unacceptable for them. Don't let that impress you too much. It merely explains why all their planes look alike: they are afraid to deviate from what they know. Too many billions of dollars, and too much fear of the unknown. Now you see why this last chapter went into so much detail and confusion. Ultimately, all you need at this stage in your course, is a reasonable number for the zero-lift drag coefficient, to use in the general Reynolds number and Mach number regime of your aircraft's cruise. Don't worry! *Remember that you have not still decided what your airplane looks like, except for the size and aspect ratio of the wings!!*

6. Your knowledge now allows you to calculate several other things. You can certainly calculate

the Power-Off Stalling Speed: that would be where the lift coefficient exceeds, say, 1.4. This poses a clear limit on how slowly your airplane can fly for a given wing loading, at a given altitude. It means that you can fly slower at low altitude where the density is high, than at high altitude where density is low. It becomes very interesting when you are trying to take off with a full payload and fuel load from an airport at a high altitude (say Denver. CO, at 5000 feet, or Leh, Ladakh, India, at 14,000 feet). Especially on a hot day.

Secondly, it also allows you to calculate the Drag Polar of your aircraft: the lift coefficient on the vertical axis, and the drag coefficient along the horizontal axis. Remember, all this is an estimate: you still don't know how big your airplane's fuselage is.

On the other hand, at this stage you are ready to calculate the size of the fuselage, allowing enough space for the fuel which you have not yet calculated. You can go ahead and experiment with that. The moment you sketch the fuselage, you can calculate its surface area, and that goes into improving your calculation of skin friction drag. Hopefully the number still fits well within the allowance that you have allotted for fuselage drag when you estimated the skin friction drag coefficient based on Benchmarking.

8.9   Example Problems

1. Example on Wing Loading and Lift to Drag ratio

   (a) An aircraft has a mass of 50,000kg and is flying straight and level at an altitude where the density is 0.5 $\frac{kg}{m^3}$. The lift coefficient is 0.3. The wing loading is 80 pounds per square foot. Find the flight speed in meters per second.

   (b) According to the computer, the speed for minimum drag at this altitude is 120 m/s. The wing spanwise efficiency factor is 0.9. The wing aspect ratio is 10. Find the zero lift drag coefficient of the aircraft.

   (c) Find the ratio of lift to total drag at the flight condition of Part (1).

2. Example 2 on Wing Loading The wing loading of an aircraft is 8000N/m 2. The value of CD0 is 0.015. The aspect ratio is 8, and the spanwise efficiency factor is 0.97. At an altitude where density is 0.4 $\frac{kg}{m^3}$, the aircraft is flying at the speed for minimum drag. Find the lift-to-drag ratio in straight and level flight

3. Example: Speed For Minimum Drag An aircraft has a zero-lift drag coefficient of 0.017, wing aspect ratio of 8, and wing loading of 7000 $\frac{N}{m^2}$. It is flying at an altitude where density is 30 percent of the sea-level standard value.

   (a) Find its speed for minimum drag.

   (b) If the sea-level static thrust of the aircraft is 30 percent of its gross weight, thrust varies proportional to density, and the stalling lift coefficient is 1.5, find the lowest speed for straight and level flight at this altitude. Check both stall and thrust criteria and show which one determines the minimum speed.

(c) If the CDmax value at transonic conditions is 0.15, and the Mach number for drag divergence is 0.8, determine whether your aircraft can reach Mach 0.85 in level flight at this altitude.

(d) Find the maximum climb rate at this altitude, at a flight speed of 200 m/s.

4. Example: Aircraft L/D An aircraft weighs 500,000 Newtons and is flying straight and level at an altitude where the density is 0.5 kg/m3. The lift coefficient is 0.3. The wing loading is 150 pounds per square foot. Find the flight speed in meters per second. This, according to the aircraft manual, is the speed for minimum drag at this weight and altitude. The wing spanwise efficiency factor is 0.9. The wing aspect ratio is 10. Find the ratio of lift to total drag.

5. Example 2: Flight at minimum drag The wing loading of an aircraft is 8000N/m 2. The value of CD0 is 0.015. The aspect ratio is 8, and the spanwise efficiency factor is 0.97. At an altitude where density is 0.4 kg/m3, the aircraft is flying at the speed for minimum drag. Find the lift-to-drag ratio in straight and level flight

6. Example: Minimum Speed An aircraft has gross weight of 2,700,000 N, aspect ratio of 10, and wing loading of 7000 $\frac{N}{m^2}$. The lift curve slope of the aircraft wings is 5.8 per radian. The zero-lift angle of attack is 0. deg. Find the minimum speed at which this aircraft can fly, at an altitude where density is 0.5 $\frac{kg}{m^3}$.

7. Example: Drama In Flight An aircraft weighing 450,000 N has a wing area of 270 square meters, and the aspect ratio is 8. The spanwise efficiency factor is 0.99. The profile drag coefficient is 0.017. Due to some dramatic events during the flight, (company VP spilled martini on the control console during surprise cockpit visit) the landing gear cannot be extended, so the ground crews at the airport are scrambling to get the runway ready with foam, fire trucks, ambulances etc. The TV anchorperson is rushing to get a final facial before reporting the upcoming landing, live. So the pilot is asked to loiter for some time, circling the area of the airport at an altitude where the density is 0.9 $\frac{kg}{m^3}$. Fuel is already pretty low. At what speed should she fly, to conserve fuel? (there is no saying how long the facial will take).

8.9.1 Example: Aerodynamics Activity

1. Download and learn to run the program, XFOIL from the Internet.

2. Select and plot the NACA 0012 airfoil shape: Y coordinate of the upper and lower surfaces as fraction of chord (y/c) vs. distance along x, the chord, as a fraction of chord (i.e., x/c).

3. Run it for the Viscous case, with a chord Reynolds Number of 100,000 and then for 1,000,000.

4. Plot lift, drad and pitching moment distributions versus angle of attack for both these Reynolds numbers. Does the program predict stall?

5. Calculate the lift curve slope that you get in each cases, using only data from the linear range.

6. For the 4 degree angle of attack case at 1 million Reynolds number, plot the pressure coefficient Cp versus chordwise distance x/c, for both the upper and lower surfaces. Comment on the difference between upper surface and lower surface.

7. Plot the same for 15 degrees. Does the Cp distribution look different? How? Is there stall occurring? What do you see in the Cp distribution that indicates stall?

8. The lift curve slope that you get (for the linear range) is the 2-D lift curve slope. Denote it by a0. Then the lift curve slope of a wing of Aspect ratio AR, made of this airfoil, is given by $a = \frac{a0}{/(1+\frac{a0e}{\pi(AR)})}$ In this equation, use the value 1 for e. It is the spanwise efficiency factor and we will assume that the designer is smart enough to get the value close to 1.

9. The wing has a span of 20m and AR of 8. You are flying it at 2000m altitude in the standard atmosphere of Earth. The airplane to which it is attached, is such that the wing loading is 800 $\frac{N}{m^2}$. Find the lift coefficient, and the induced drag coefficient.

10. Now find the angle of attack that gives this wing lift coefficient. The wing is rectangular (so chord is the same everywhere). You know the viscous drag coefficient of the airfoil from the XFOIL program for the angle of attack of the wing. So now find the total drag, and the lift to drag ratio.

11. Find the speed for minimum speed at this altitude for this airplane.

12. An aircraft weighs 500,000 Newtons and is flying straight and level at an altitude where the density is 0.5 $\frac{kg}{m^3}$. The lift coefficient is 0.3. The wing loading is 150 pounds per square foot. Find the flight speed in meters per second. (NOTE: Wing Loading is W/S, and has units of $\frac{N}{m^2}$) This, according to the aircraft manual, is the speed for minimum drag at this weight and altitude. The wing spanwise efficiency factor is 0.9. The wing aspect ratio is 10. Find the ratio of lift to total drag.

13. A wing has an aspect ratio of 8, spanwise efficiency factor of 0.9 and supports an aircraft of W/S = 3000 $\frac{N}{m^2}$. The lift-independent or profile drag coefficient of the aircraft is 0.02. It is flying at an altitude where density is 0.9 $\frac{kg}{m^3}$ Find the speed for minimum drag. If the lift-curve slope of the wing is 5.3 per radian, find the effective angle of attack of the wing at this condition (note: the effective angle of attack is the only kind you can find at this stage).

8.10   Design Update: Steps in Designing a Hydrogen Supersonic Airliner

It is time to do some serious calculations on the hydrogen-powered aircraft. We already know the payload and the Take-Off Weight. We will assume that the wing loading is 100 lbs per sq.

foot in a "typical cruise" condition.

1. Calculate the wing area in cruise.

2. Given that the wing span is limited to 200 feet, find the aspect ratio.

3. Give a name to your airplane: it now has a geometry, weight, etc.

4. Construct an Excel file with all the values you know for the airplane, and another file which gives you the temperature, pressure and density if you type an altitude in meters in the Standard Earth Atmosphere.

5. Given that the profile drag coefficient is 0.012, find the speed for minimum drag at each altitude, from sea-level to 20,000 meters, in steps of 2000 meters. The spanwise efficiency factor is 0.99.

6. Find the drag at each altitude (this is the thrust required)

7. Plot the results of steps e and f, with the altitude on the horizontal axis. Divide each quantity by its value at sea level, so that the magnitude is visible on the same plot. Give these sea-level values on the plots.

Chapter 9

Propulsion

Aerospace Propulsion! Conjures visions of zooming jet engines, glowing jet exhausts shaking the neighborhood with the thunder of sheer power. Propulsion is an incredibly complex discipline that brings in our best ideas from every field of science and technology. Solid propellants are prepared by methods that would make a gourmet chef blush in envy: it is literally cooking, with organic and inorganic materials that can burn fiercely or explode. Liquid propellants can be extremely toxic, borderline explosive, or deadly cryogenic. Combustion chemistry can involve reaction schemes with over 2200 different species formed as intermediate products, in solid, liquid and gas phases, with over 11,000 different reaction steps all occurring simultaneously. Computing the final temperature and composition of a jet engine combustor exit flow challenges the fastest of computers even with the best of mathematics in algorithms for solving simultaneous differential equations. Oh, yes, my MSAE and PhD were in Propulsion/Combustion. I did get to learn from some legendary Rocket Scientists.

Propulsion is the science and engineering of systems to move (propel) aircraft, missiles and spacecraft to their destinations. Vehicle propulsion is defined [123] as the action or process of imparting motion to a vehicle by means of a force. Typically the force is generated by changing the momentum of some fluid such as air, by adding energy to the fluid. Propelling a vehicle involves generating a force in order to accelerate a vehicle from one state of momentum to another, or to balance other forces to maintain a given state in equilibrium. This is the field of rocket engines, jet engines, internal combustion engines and pulsed detonation engines, but it also deals with ion engines, solar sails, nuclear engines, and even matter-antimatter engines. While the actual engines appear to be immensely complex, the underlying design and operation principles are the basic laws of physics and thermodynamics.

9.0.1 Thrust and Power

Two basic propulsion concepts are Thrust and Power. Thrust is the force that the propulsion system exerts on the vehicle, measured in Newtons (N), or pounds force (lbf). Power is work done per unit time, measured in Watts or foot-pounds per second. Values range from the micro-

Newtons of some engines used in spacecraft propulsion, to millions of Newtons for large launch vehicles.

But we are not here to dazzle ourselves with complexity. We are here to cut through all that, and focus on what we need. At this stage, as we design a new aircraft, we are not going to worry about designing a new engine as well. We will learn to select a suitable engine from those already available, or at least settle on one that could be developed with reasonable effort and moderate risk in the next ten years, given a few different engines and technologies that exist today. So let us see how to do this.

How much thrust do we need? Obviously that translates to the size of the engine. If it is a rocket that goes up vertically, the thrust must be at least slightly (say 10 percent?) higher than the weight. But flying with aerodynamics is much smarter: once we start flying, the job of balancing the weight is done by aerodynamic lift. We only have to make sure that that aircraft is moving fast enough to generate enough lift.

As a rule of thumb, maximum takeoff thrust available for a vehicle that flies using aerodynamic lift, must be around 30% of the gross weight of the aircraft. We will see why, presently. For subsonic/transonic airliners, takeoff is the most demanding thrust condition, so it drives the selection of the engines. Airliners must be able to take off with one engine failed at the worst time possible: just as the aircraft is lifting off the runway, with not enough runway left ahead to land right away. Modern airliners use 2, 3 or 4 engines, trending towards using only two if possible.

A note on this maximum thrust value. It is not always true that takeoff is the most demanding condition. Imagine a supersonic airliner or a fighter plane, that must fly very fast at a high altitude where the air is much less dense. The Power needed, is that which is enough to overcome the drag. The drag is proportional to the square of speed. The power is the drag times the speed (point of application of a force, being moved through a distance, per unit time). So Thrust Required is proportional the square of speed, and the Power is proportional to the cube of speed. All right so far.

But here comes a nasty fact: thrust is generated by increasing the momentum of air. Air density at say, 50,000 feet altitude is very low compared to that at sea-level. So the maximum thrust that can be generated, may be quite low compared to what the engine can do, standing still at sea-level (called Sea Level Static Thrust). For the supersonic aircraft, the maximum thrust will be dictated by the thrust required at altitude. In other words, you may end up with an aircraft that has a large amount of excess thrust at sea-level. An over-sized engine. This problem requires innovative solutions. Of course, reducing supersonic drag would help. Or, ways to gulp in air over a larger frontal area without increasing drag too much. Or something else that you may invent. No birds fly supersonic, so we cannot copy them.

9.1   How Jet Engines Work

Jet engines work using the Gas Turbine Cycle, a process which consists of the following four steps: compression, heat addition, expansion, and cooling.

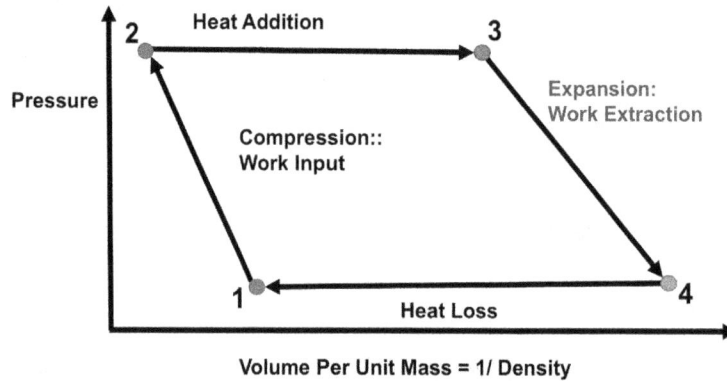

Volume Per Unit Mass = 1/ Density

1. Compression Air enters the engine at P_1, and is compressed to a very high pressure P_2. In fact this is the highest pressure in the engine. This is done by doing work on the air using the rotating blades of a fan and a compressor. The temperature also rises as the air is compressed.

$$\frac{P_2}{P_1} = \left(\frac{T_2}{T_1}\right)^{\frac{\gamma}{\gamma-1}} \text{ where } \gamma = 1.4 \text{ so that } \frac{\gamma}{\gamma-1} = 3.5$$

2. Heat Addition Heat is added to high-pressure air by burning fuel. Pressure remains constant, and temperature rises to its highest point in the engine. This temperature is called the turbine inlet temperature, as it will enter the turbine after this point. This quantity can be controlled by varying the amount of fuel added. In modern military engines, this temperature is greater than 2000 Kelvin.

3. Expansion and Work Extraction Hot, high pressure air is blown out through a turbine, and then accelerated through a nozzle. The air moving through the turbine blades forces it to spin. The turbine is connected to the compressor with a shaft, and the turbine drives the compressor as air flows through it. This takes work out of the air, lowering its pressure and temperature. The air then blows out of the nozzle as a jet, with a velocity u_e and a static pressure equal to ambient air pressure, or $p_e = p_a$

4. Cooling: To complete the cycle, we must consider the cooling of air (constant-temperature heat release) in the atmosphere before the next jet aircraft comes along and gulps it in. However, since we don't directly pay for this step inside the engine, we don't worry about it much. Isn't it a growing worry as it is part of the contribution of aerial vehicles to atmospheric heating, that is known as Global Warming? Well... the conversion of hydrocarbon fuels to carbon dioxide is much more worrisome because the released carbon dioxide remains in the atmosphere and absorbs solar radiation. The heat release from air is quite small by comparison as engines become more efficient.

9.2   Thrust of a Jet Engine

Thrust is calculated using Newton's Second Law of Motion.

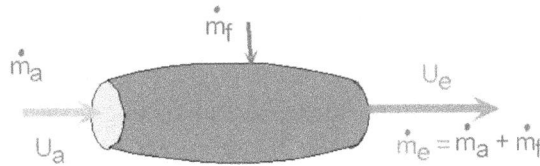

Figure. 9.1: Basic jet engine.

Let's say that the speed of air entering the engine is u_a. The engine gulps $\dot{m}_a$ kg/s of air, moving at u_a with respect to the engine. It adds $\dot{m}_f$ kg of fuel per second, which has about zero velocity with respect to the engine. All this mass then blows out of the exhaust at u_e m/s.

$$(\dot{m}_a + \dot{m}_f)u_e - \dot{m}_a u_a - \dot{m}_f(0)$$

Defining fuel-to-air ratio as: $f \equiv \frac{\dot{m}_f}{\dot{m}_a}$
Rate of change of momentum is: $\dot{m}_a[(1+f)u_e - u_a]$
This must be equal to the thrust. $\tau = \dot{m}_a[(1+f)u_e - u_a]$

The function of a jet engine is to increase the momentum of the fluid passing through it.

Pressure Thrust: The general thrust equation in the previous section left out a pressure term. A more accurate equation of thrust would be:

$$\tau = \dot{m}_a[(1+f)u_e - u_a] + (p_e - p_a)A_e$$

where A_e is the area of the engine's nozzle exit, p_e is the pressure of the gas exiting the engine, and p_a is the ambient pressure. The extra term could be called pressure thrust, which is the force due to the pressure difference between the exhaust plane and the outside, acting on the exhaust area. Ideally, a nozzle should expand the flow such that the exit pressure is equal to the ambient pressure, or $p_e = p_a$. Most well-designed nozzles for flight inside the atmosphere come pretty close to this, and in many cases this extra term can be dropped. In addition, it is true that if the flow coming out of the nozzle is subsonic, for the speed of sound based on the temperature of the exhaust (which is pretty hot) then the pressure at the exit of the nozzle will adjust to the outside pressure, with an appropriate exit velocity. However if the flow is supersonic, such as the exhaust of a fighter plane on afterburners, or a rocket nozzle, then the flow inside cannot "know" the pressure outside, so it comes out at whatever speed and pressure are dictated by the geometry of the nozzle.

In the case of rockets which operate at very high altitudes or in the vacuum of Space, the pressure thrust term is essential. The nozzle can never be made large enough to expand the flow all the way to the outside pressure which may be zero! In fact the nozzles of large rocket engines such as those of the Space Shuttle Main Engines, are designed to work optimally at some intermediate altitude. At liftoff you can see, if you look very carefully, the flow coming out of the nozzle converging a bit. This is because the nozzle expands the flow to a pressure that is lower than the outside atmospheric pressure! Yes, all this is way beyond such an introductory textbook, but there is no harm in your learning how simple these concepts are. When you next see them in the Gas Dynamics course under the heading of Converging-Diverging Nozzles, your classmates who do not read this book are going to be so immersed in feeling so sorry for themselves at seeing the beautiful logic of that subject (*"Do I really HAVE to MEMORIZE these EQUATIONS???"*) , that they will not realize the simplicity of the concept.

9.3   Turbojets, Turboprops, Turbofans

Turbojets, turboprops, and turbofans are three different types of air-breathing jet engines that employ the use of a turbine (the "turbo" prefix in each of these words means a turbine is involved). However, some have additional turbines that extract energy out of the flow to do different things.

1. Turbojets In turbojets, the turbine only takes energy out of the flow for the compressor and some auxiliary services, but little else. Most of the hot gas expanded during the burning phase is rushed out the nozzle as a fast and generally quite hot, jet of air. The airliner engines of the late 1960s and early 1970s used to roar with the noise of these jet exhausts. Jet Noise was a popular (well-funded) field of research.

2. Turbofans In turbofans, there is another turbine in the engine that takes energy out of the flow to turn a fan at the front of an engine. This fan accelerates air outside the core of the engine, which adds to the thrust. The air accelerated by the fan is called bypass air. The ratio of the mass flow rate of air through the fan to the mass flow rate of air through the core of the engine is called the bypass ratio, denoted by β.

$$\tau = \beta \dot{m}_{aH}(u_{eC} - u_a) + \dot{m}_{aH}[(1+f)u_{eH} - u_a]$$

In the above, the H subscript denotes the flow that goes through the hot part of the engine, meaning the compressor, burner and turbine. The C subscript denotes the (relatively) cold part of the flow, which goes only through the fan and the fan nozzle. The letter a refers to the ambient conditions, meaning the outside air, and the e subscript refers to the exit. The letter f refers to the fuel to air ratio, usually defined as the ratio of the fuel mass flow rate to the mass flow rate of air through the hot part of the engine.

Turbofans are very efficient because the fan moves a lot of air at a slow velocity. For this reason, most commercial airliners use turbofans. The bypass ratio for most of these engines is usually between 4 to 8.

3. Turboprops In turboprops, a turbine takes a lot of energy out of the flow, and is connected to a gearbox at the front of the engine, which is then connected to a propeller. This propeller is what gives the engine most of its thrust; only 15 - 20% comes from the jet. The

gearbox is necessary to reduce the rotational speed of the engine down to propeller speeds.

4. Ramjets A ramjet is an air-breathing jet engine that has less moving parts than a turbojet. The air is compressed without a compressor, relying solely on the geometry of the engine inlet to compress the air. This relieves the need for a turbine. As a result, the combustor can heat the gas to higher temperatures. There are no turbine blades on which the flow can stagnate, generating temperatures above the melting point of the blades. The ramjet contains less complicated machinery, which is good from a manufacturing standpoint. The thrust equation for ramjets is the same as the general equation for thrust out of a jet engine:

$$\tau = \dot{m}_a[(1+f)u_e - u] + (p_e - p_a)A_e$$

Assuming an ideal nozzle, where $p_e = p_a$,

$$\tau = \dot{m}_a[(1+f)u_e - u]$$

A huge problem with ramjets is that they cannot generate thrust when they are not moving. Because a ramjet has no compressor, it has no way of sucking air into the engine. There are three solutions to this. The first is that the ramjet is integrated with another engine. It may be a turbojet/turbofan engine such as what the SR-71 engines used. At high Mach numbers, it operated as a ramjet. Or it may be a rocket engine that starts the vehicle and takes it up to a high Mach number before the ramjet starts up. In some cases, the air intake for the ramjet is the space where the solid fuel used to be, that got burned in the rocket. A third is that the ramjet is simply launched from a vehicle that is already moving at near sonic speed. Although a ramjet can "start" at about Mach 0.6, the thrust does not really become significant until well past Mach 1. The best operating regime for a ramjet is around Mach 2 to Mach 2.5.

9.4 Rockets

Rockets do not take in any air, and must carry oxidizer as well as fuel. Rockets generate thrust by accelerating this fuel-oxidizer mix out the back of the engine. The thrust equation for a rocket is:

$$\tau = \dot{m}_e u_e + (p_e - p_a)A_e$$

The pressure term in the thrust for a rocket engine is more important than in the other engines examined. In most other cases, the engine can be designed so that the exit pressure is equal to the atmospheric pressure, thus eliminating the pressure term. However this is more difficult with rockets because of two reasons. First, rockets tend to work in low-pressure environments like high altitude, or zero-pressure environments like space. Thus it is harder for the nozzle to expand the flow to the correct conditions. Also, rockets often change altitude very quickly. This rapid change in the ambient pressure surrounding the rocket makes it difficult for a single nozzle design to expand the flow to ideal conditions at each altitude.

Rockets have a measure called the effective exhaust velocity, c_e. This term can be found by adding both types of thrust in the rocket thrust equation and dividing by the total mass flow rate of gas exiting the nozzle:

$$\tau = \dot{m}_e u_e + (p_e - p_a)A_e = \dot{m}_e c_e$$

$$c_e = u_e + \frac{(p_e - p_a)A_e}{\dot{m}_e}$$

Specific Impulse: The concept of equivalent exhaust velocity leads directly the very important idea of Specific Impulse. The Specific Impulse is the equivalent exhaust velocity, divided by the standard acceleration value of 9.8 m/s^2. Thus it has units of seconds. NOTE: We did not say, divide by the acceleration due to gravity! Just the value 9.8 m/s^2 which is easy to remember because it happens to be the approximate value of acceleration due to gravity at the Earth's surface. Specific Impulse is a measure of the goodness (or Ooomph!) of a propulsive system. Some Space agencies express the value without dividing by the constant, which leaves the value with velocity units. A recipe for disaster: people always get confused between the British and Metric systems of units, but a second is a second, in all systems.

9.5   Efficiency

1. **Thermal Efficiency**: Thermodynamics shows that the efficiency in converting heat to work is highest when the heat is added at the highest possible pressure.

$$\eta_{thermal} = 1 - (\frac{1}{P_r})^{(\gamma-1)/\gamma} \tag{9.1}$$

    The compressor pressure ratio, denoted as P_r, is the ratio of pressures after and before the compressor. Modern turbine engines have high compressor pressure ratios, meaning they compress the incoming air to many times its original pressure. Note that even when P_r is 40, the thermal efficiency is only 0.65, or 65%. As with most efficiencies, thermal and propulsive efficiencies can only have values between 0 and 1.

2. **Propulsive Efficiency**: The propulsive efficiency measures the efficiency in converting the kinetic energy of the fluid (air) to thrust.

$$\eta_{propulsive} = \frac{2}{1 + \frac{u_e}{u}} \tag{9.2}$$

    The exit velocity u_e will always be greater than the flight velocity u if thrust is created. Taking air in at a certain velocity, the engine must shoot it out at a greater velocity for there to be a change of momentum in the fluid. The thermal efficiency is maximized by driving the exhaust velocity as close as possible to the flight speed u. This is done by using high bypass fans, or turboprops. These devices move lots of air slowly, so that the exit velocity u_e is as close to u as possible.

3. **Thrust Specific Fuel Consumption:** This quantity, called TSFC or tsfc, is the thrust divided by the weight of propellant expended, per *hour*. Note the terms Weight and Hour. These are departures from what we might expect in the SI system, and working with mass, but this is the traditional way of expression thrust-specific fuel consumption.

   Low TSFC is good. Large commercial aircraft turbofan engines with high bypass rations (like 12) offer TSFC as low as 0.33. A jet engine with full afterburners on, may have 2.6. Engines without afterburners will be in the range of 0.8 to 1 or thereabouts. The above engine is a moderate-bypass turbofan, perhaps powering a smaller aircraft such as a business jet.

4. **Thrust Lapse Rate:** As we go up in the atmosphere, the density decreases. So, for a given flight speed, thrust decreases as altitude increases. This rate of decrease is called the thrust lapse rate. In the absence of any data on this lapse, it is reasonable to assume that thrust is proportional to the density of air. Empirical relations have been developed for large turbofan engines of the type employed on modern commercial transport aircraft. We will not list those here, but you can find them easily for given engines, from the manufacturers. Below 5000 m, it can be assumed that thrust varies linearly, or scales with density. The variation of thrust with speed is milder than the variation with altitude. The expression is not monotonic (i.e., it does not just keep increasing, or keep decreasing, with increasing Mach number), so it is not attempted here.

5. Example: Thrust Calculation Calculate the thrust and specific fuel consumption of the engine with the following conditions:

   Hot air mass flow rate $= 100$ kg/s

   Flight speed $= 200$ m/s

   Fuel/air ratio $= 0.015$

   Hot exhaust velocity $= 800$ m/s

   Fan exhaust velocity $= 250$ m/s

   Bypass ratio $= 6$

Solution:
Because there is a bypass ratio of 6, we know this is a turbofan engine. Using the equation provided in the turbofan section above,

$$\tau = \beta \dot{m_{aH}}(u_{eC} - u_a) + \dot{m_{aH}}[(1+f)u_{eH} - u_a]$$

$$\tau = (6)(100 \text{ m/s})(250 \text{ m/s} - 200 \text{ m/s})$$

$$+(100 \text{ kg/s})[(1+0.015)800 \text{ m/s} - 200 \text{ m/s}]$$

$$\tau = 91200 \text{ N}$$

TSFC:

$$\text{TSFC} = \frac{\dot{m}_f}{\tau}$$

$$\text{TSFC} = \frac{(0.015)(100 \text{ kg/s})(9.8((3600)}{91200 \text{ N}}$$
$$\text{TSFC} = 0.58$$

(per hour).

This is TSFC in units of fuel weight per hour, per unit thrust. Some Implications:
Assume thrust to weight ratio of the engine is roughly 4.5, so that the weight of the engine
is 20335 N. Assume that the engine is installed using the 30% of TOW guideline. TOW =
305333 N. Engine weight fraction = 20335/305333 = 6.65%.

If L/D = 12, thrust needed at flight altitude is only 25444 N. Can fly at altitude where
density is about 0.28 of sea level.

9.6   Going Beyond

Textbooks on Propulsion include [17–28, 124]. Other books on specialized topics related to
propulsion include [125–129]

9.7   Design Update With Engines

In brief, we will

1. Select engines based on thrust and 1-engine-out criteria.

2. Find thrust-specific fuel consumption.

3. Estimate thrust variation with altitude.

OK! Let us catch up with our Design process. This is the point in the course where people start
feeling sorry for themselves. So many concepts! So many EQUATIONS! Let us pause and update
our design process and knowledge. What did we do so far?

1. We started with a Requirements Definition.

2. We calculated the TOW using a Payload Fraction estimate and the required payload.

3. Benchmarking also gave us all the other fractions as estimates. We must remember to
check that we come in under the allowance for each of these, or the design will not *close*,
as the Insiders call it. Two big red flags are a fuel fraction getting past about 40 percent
(Ok, 42 percent may be OK) and the structure fraction remaining, going below 28 percent.

4. We used benchmarking to pick a value for W/S, and from that we found S. With the span decided by practical constraints to decide the span, and from that we found aspect ratio. This allowed us to calculate the lift coefficient and induced drag coefficient for any speed and altitude, given the wing loading. We estimated the zero-lift drag coefficient using the ideas of Reynolds Number and Mach Number. From all this we were able to construct the Drag Polar of the aircraft.

5. Now comes our new-found knowledge of Propulsion. We start with an assumption on the amount of thrust needed. The beauty of aerodynamic lift is that you can get off the ground with thrust not more than about 10 percent of weight. But we typically install as much thrust as 1/3 of the weight at takeoff, to survive things like engine failure.

6. At this stage, we merely SELECT an engine, so that its parameters are known to us.

7. The thrust-to-weight ratio of the engine is a matter of technology, so we can easily verify that the engine weight is indeed well within the allowance that we allotted at the start of the process.

8. The variation of thrust with altitude and flight speed can be obtained, knowing the engine. The usual bad news is that thrust decreases as altitude increases, approximately as the density decreases. This allows us to make sure that at the cruise altitude and Mach number, we have enough thrust from our engines, operating at about 80 percent of their maximum available thrust, to overcome the drag. If that does not work, we must increase the engine size. In general the thrust at a given altitude is highest at zero speed (static thrust), but as we get into the transonic and supersonic range, the thrust does actually go up, and can be higher than the static thrust at that altitude.

9. The thrust-specific fuel consumption is also something that we can find from the engine data.

10. Knowing the above, we should be able to calculate the amount of fuel consumed. But for that we must learn more about Performance.

9.8 Example Problems

9.8.1 Rocket Problems

1. A rocket propulsion system has a specific impulse of 390 seconds. Find the equivalent exhaust velocity in meters/ second. Find the mass ratio (initial mass divided by the mass left after the propellant is all gone) to achieve a velocity increment of 1500 m/s.)

2. A rocket is accelerating straight upwards. The weight of the rocket is 5,000,000 N, the velocity U is 900m/s, and the drag is given by $0.02U^2$. The rocket is producing a thrust of 7,500,000 N. Find the velocity and the distance traveled, 0.1 second later, assuming that the weight does not change during that time.

3. A rocket propulsion system has a specific impulse of 390 seconds. Find the equivalent exhaust velocity in meters/ second. Find the mass ratio (initial mass divided by the mass left after the propellant is all gone) to achieve a velocity increment of 1500 m/s.

4. A rocket propulsion system has a specific impulse of 390 seconds. Find the equivalent exhaust velocity in meters/ second. Find the mass ratio (initial mass divided by the mass left after the propellant is all gone) to achieve a velocity increment of 1500 m/s.

5. A rocket engine has an exhaust mass flow rate of 500 kg/s, exhaust velocity of 3500 m/s and a pressure thrust of 300,000N. Find the effective exhaust velocity and the specific impulse.

9.8.2   Jet Engine Problems

1. An aircraft is to carry a payload of 30,000 kg over a range of 2000km, cruising at 13,000 meters standard altitude, at 250m/s. Benchmarking shows that aircraft in this category have a payload fraction of 0.35, so that we can try for a payload fraction of 0.29 with our advanced technology. It is a rule of thumb that the installed engine thrust available at takeoff should be 30% of the takeoff gross weight. What is this value of installed thrust for this aircraft?

2. A large commercial aircraft which you are designing can achieve a payload fraction of 20%. The payload is 60000kg. As a general rule, maximum thrust available at takeoff should be 30% of takeoff weight. How much thrust should the engines have at takeoff at sea-level standard conditions? Express in Newtons, and in lbf.

3. You are considering the design of a supersonic airliner. Data on designs of this type show that a payload fraction of 15% may be feasible. You have to have a payload of 120,000 lbs. It is also known that the engines must be big enough to generate a maximum thrust of 30% of the takeoff weight. Further, engine design is so advanced today that the thrust of an engine is about 4.5 times the weight of the engine. Find the engine weight.

4. A jet engine takes in air at 50,000 N/m2 and compresses it to 2,000,000 N/m2 before adding heat to it. Calculate the ideal thermal efficiency of this engine.

5. A jet engine has a bypass ratio of 12, fuel-to-core air ratio of 0.02, cold exhaust velocity of 300 m/s and hot exhaust velocity of 600 m/s, at static conditions. Find the fuel mass flow rate needed to generate 200,000 N of thrust, and also find the thrust-specific fuel consumption in the usual units for TSFC.

6. A jet engine flying at 100 m/s takes in 100 kg/s of air. The exhaust velocity is 500 m/s. The mass of fuel being consumed per second is 2 kg/s. Find the thrust of the engine.

7. A turbofan engine has a cold mass flow rate of 400 kg/sec., bypass ratio of 4, and the fuel flow rate is 1.5kg/s. The hot exhaust comes out at 1100 m/s, the cold exhaust comes out at 320 m/s, and the flight speed is 200 m/s. Find the thrust in Newtons and the thrust-specific fuel consumption (in units of Per Hour).

8. A turbofan engine has the following parameters at sea-level under static conditions: Cold air mass flow rate is 100 kg/s, fuel-air ratio 2%,. Bypass ratio is 8. Hot exhaust velocity is 800 m/s. Cold exhaust velocity is 300 m/s. Maximum turbine inlet temperature is 2000K. Sea level standard temperature is 288K, and compressor efficiency is estimated at 83%. What is the maximum TOW of an aircraft which you would power with two of these engines?

9. A new jet engine is operating in steady cruise. The incoming mass flow rate is 200 kg/s, at a stagnation temperature of 400K, the flight speed is 100 m/s, the Overall Pressure Ratio is 35, and the highest temperature reached at the entrance to the turbine is 2000K. The fuel to air ratio is 0.019. The jet exhaust comes out at 800 m/s. Find the thrust. Also, find the thrust-specific fuel consumption.

Chapter 10

Performance

Aircraft performance is the study of how well an aircraft achieves specific goals. These goals depend on the specific mission of the aircraft in question. While everyone would like the best of everything, one can do better by focusing on what is really valuable. Transport aircraft need a lot of range, while patrol aircraft need long endurance. Customers of transport aircraft may value a short-field takeoff and landing capacity even more than range. Fighter aircraft prize a small turning radius so as to have the edge in a dogfight, but interceptor aircraft also require good climb performance to quickly meet threats. This chapter explores some of the most common performance parameters, and shows how to evaluate them given some basic information about the aircraft. Under performance, we think about:

1. How long can the aircraft stay in the air (endurance)?

2. How far can it fly (range)?

3. How much payload can it carry on a given mission?

4. How long will it take to reach altitude (climb performance)?

5. What is its maximum speed?

6. How long a runway length does it need to take off?

7. How long a runway does it need to land?

8. How quickly can it turn, pitch, and roll?

9. What is its minimum turning radius?

10.1 Basic Performance Parameters

In this section, we discuss basic performance parameters. These are either simple enough not to require an entire section, or necessary for understanding more complex parameters later in the chapter.

1. **Wing Loading:** The wing loading of an aircraft is given by the total weight of the aircraft divided by the planform area of the wing. The Boeing 747 has a wing loading of about 150 lbf/ft^2 (psf) or 7180 N/m^2. In order to grasp the magnitude of this feat, try to visualize a square foot of metal - say a bathroom scale or a floor tile? Then imagine putting 150 pounds of weight on it (a human standing there) and trying to make it fly.

2. **Excess Thrust:** For an aircraft to fly straight and level, its thrust must equal its drag. However, if an aircraft's engines are producing more thrust than is needed to fly, it is said to have excess thrust. Excess thrust is simply the difference between thrust and drag.

$$\text{Excess Thrust} = T - D$$

3. **Excess Power** The excess power is simply the excess thrust multiplied by the velocity at which the aircraft is moving.

$$\text{Excess Power} = U_\infty(T - D)$$

4. **Steady, Unaccelerated Climb:** Steady, unaccelerated climb refers to climb at a constant velocity U_∞ at some nonzero angle θ to the horizontal. This angle θ is called the climb angle. Usually the most important parameter involved with climb performance is rate of climb, denoted R_c, which is the rate of change in altitude. It is the vertical speed of an aircraft, and is most commonly expressed in feet per minute, even in metric countries.

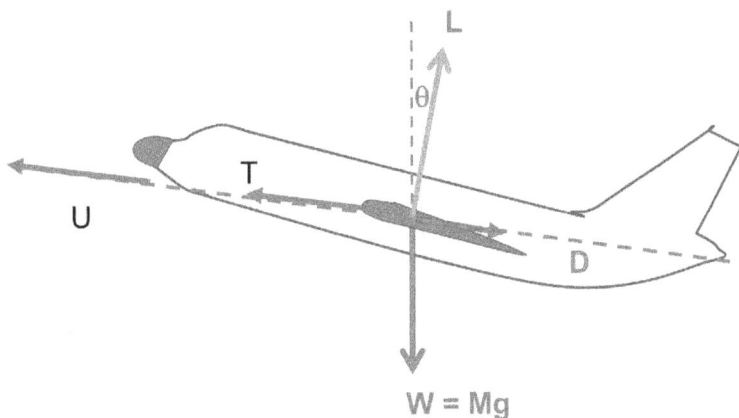

Figure. 10.1: Aircraft in steady climb.

From Figure 10.1, we can derive the equation of motion along the freestream:

$$T - D - W\sin\theta = 0$$

and the equation of motion perpendicular to the freestream:

$$L - W\cos\theta = 0$$

Solving for $\sin\theta$ in the equation of motion along the freestream gives:

$$\sin\theta = \frac{T-D}{W}$$

The aircraft is climbing with a speed U_∞ at an angle θ to the horizontal. This can be broken into components, as shown below.

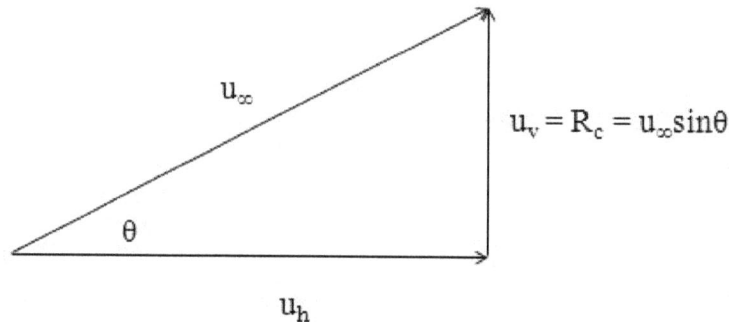

Figure. 10.2: Decomposing the velocity into horizontal and vertical components

5. **Rate of climb:** From Figure 10.2, we can see that the rate of climb, R_c, is equal to the vertical component of the velocity, U_v, which is equal to $U_\infty\sin\theta$.

$$R_c = U_\infty\sin\theta$$

Inserting the equation for $\sin\theta$ derived above,

$$R_c = U_\infty\frac{T-D}{W}$$

The term $U_\infty(T-D)$ is the term for excess power. Thus, the rate of climb can be expressed as excess power over weight.

$$R_c = \frac{\text{Excess Power}}{W}$$

Because of this relationship between excess power and rate of climb, it can be said that excess power determines the rate of climb. Similarly, excess thrust determines the angle of climb, θ. This can be seen if the expression for climb angle is slightly modified:

$$\sin\theta = \frac{T-D}{W} = \frac{\text{Excess Thrust}}{W}$$

Example: Steady Climb

An aircraft has a L/D of 15 and its thrust is 105% of the drag. Find the climb angle, assuming constant flight speed.

L/D = 15
T/D = 1.05

$$\tan\theta = (T-D)/L = (T/D-1)/(L/D) = 0.05/15$$

$$= 0.0033$$

$$\sim 0.0033 * 180/\pi \text{ degrees}$$

$$\sim 0.2 \text{ degrees}$$

If speed is 200 m/s, climb rate is 0.0033*200 = 0.66 m/s ~ 120 feet per minute. Very low.
Typical climb rate is ~ 500 fpm.
Question: what is the thrust needed, as % of drag, for a 500 fpm climb?

---

6. **Gliding Flight:** Gliding flight, or unpowered flight, is flight with no thrust. Although there are some aircraft specifically built for gliding, any aircraft can become a glider if the engine fails or is shut off. The glide angle is the angle between the horizontal and the flight path of the aircraft, and is denoted by θ. The equilibrium equations for the figure above are shown below:

$$D = W\sin\theta$$

$$L = W\cos\theta$$

$$\frac{D}{L} = \tan\theta = \frac{1}{L/D}$$

The lift-to-drag ratio determines the glide angle θ. The higher the lift-to-drag ratio, the smaller the glide angle.

$$\tan\theta_{min} = \frac{1}{(L/D)_{max}}$$

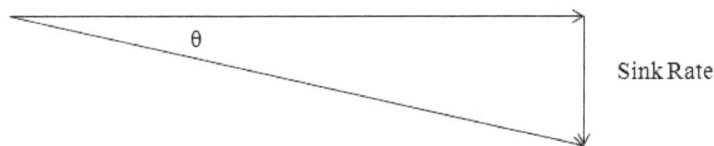

Figure. 10.3: Sink Rate

7. **Sink Rate:** The sink rate, or rate of descent, is the same as the rate of climb but in the opposite direction. It is the vertical component of the aircraft's velocity.

$$\text{Sink Rate} = U_\infty \sin\theta$$

This can also be expressed using the drag and weight of the aircraft:

$$\text{Sink Rate} = U_\infty \frac{D}{W}$$

8. **Minimum Rate of Descent:** Minimum rate of descent occurs when the glide angle is smallest. Because the smallest glide angle is achieved with the highest lift-to-drag ratio, the minimum sink rate occurs when the L/D ratio is highest.

It is possible to find the glide velocity U_∞ for a given θ.

$$L = \frac{1}{2}\rho U_\infty^2 S C_L$$

$$\frac{1}{2}\rho U_\infty^2 S C_L = W\cos\theta$$

$$U_\infty = \sqrt{\frac{2\cos\theta}{\rho C_L}\frac{W}{S}}$$

Example: Sink Rate
An aircraft is coming in on final landing approach. The pilot has aligned the craft with the centerline of the runway, and trims for constant speed of 65 mph. If the aircraft weight is 2400 lb and drag is 220 lb, what should the thrust be, to achieve a constant sink rate of 500 ft/min?

Solution:

$$W = 2400 \text{ lb}$$

$$D = 220 \text{ lb}$$

$$U_\infty = 65\ \frac{\text{mi}}{\text{hr}}\cdot\frac{1\text{ hr}}{60\text{ min}}\cdot\frac{5280\text{ ft}}{1\text{ mi}} = 5720 \text{ ft/min}$$

$$\text{Sink Rate} = U_\infty \sin\theta$$

Unlike a problem with purely gliding flight, this one includes thrust:

$$W\sin\theta + T = D$$

$$\sin\theta = \frac{D-T}{W}$$

$$\text{Sink Rate} = U_\infty \frac{D-T}{W}$$

$$T = D - W\frac{\text{Sink Rate}}{U_\infty}$$

$$T = 220 - (2400)\left(\frac{500}{5720}\right)$$

The thrust must be 10 lb to achieve a constant sink rate of 500 ft/min.

---

9. **Accelerated Climb:** The sections on steady climb and gliding flight both focused on equilibrium situations, with no acceleration. Now we look at accelerated climb, but will use energy methods instead of dealing with free body diagrams and Newton's second law. The first concept that must be understood is energy height. The total energy of an aircraft flying at some altitude h is given by the sum of its potential and kinetic energies.

$$\text{Total aircraft energy} = mgh + \frac{1}{2}mU_\infty^2$$

This fixed amount of energy can be "spent" in any combination of potential energy or kinetic energy. The total aircraft energy could be turned into all potential energy, or all kinetic energy. An aircraft flying at an altitude h and some velocity U_∞ could reduce its altitude to almost zero, which would convert all the potential energy into kinetic energy, increasing the speed of the aircraft. Alternatively, the aircraft could convert all of its kinetic energy into potential energy, which would increase the altitude of the aircraft (though it would have no velocity!). This last idea illustrates the idea of energy height, H_e. Energy height is simply the total aircraft energy divided by the weight of the aircraft:

$$H_e = \frac{mgh + \frac{1}{2}mU_\infty^2}{mg}$$

$$H_e = h + \frac{U_\infty^2}{2g}$$

Energy height has units of height and represents the altitude an aircraft could reach if all of its kinetic energy were converted to potential energy.

10.2   Level Turn Performance

A level turn is a turn in which the altitude of the aircraft remains constant. In order to perform a level turn, an aircraft rolls an angle ϕ. This causes the lift vector to rotate by ϕ, as shown in Figure 10.4. The horizontal component of the rotated lift vector is what causes the aircraft to turn.

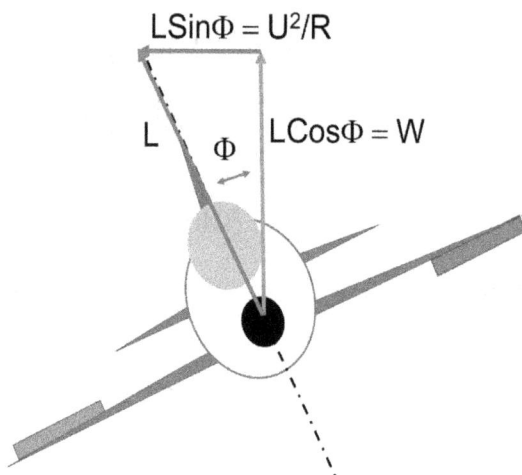

Figure. 10.4: Aircraft performing a level turn.

Because the turn is level, and there is no change in altitude, there can be no net force in the vertical direction. This leads to the following relation between lift and weight:

$$L\cos\phi = W$$

In straight and level flight, the lift must equal the weight for the aircraft to remain in equilibrium. For any angle ϕ greater than $0°$ and less than $90°$, the value $\cos\phi$ will be less than one. Therefore lift must be greater than the weight if the aircraft is to remain at a level altitude during its turn.

The centripetal force experienced by the aircraft is equal to the horizontal component of the lift vector.

$$m\frac{U_\infty^2}{R_t} = L\sin\phi$$

In the equation above R_t is the turn radius. Thus the equation for turn radius is

$$R_t = \frac{m}{L}\frac{U_\infty^2}{\sin\phi}$$

10.2.1   Load Factor

The load factor n is defined as the ratio of lift to weight. It is of great importance when looking at turn performance.

$$n \equiv \frac{L}{W}$$

We have shown that for a level turn,

$$L\cos\phi = W$$

thus

$$n = \frac{L}{W} = \frac{1}{\cos\phi}$$

For a level turn, the load factor and the roll angle are directly related. The load factor is often expressed in terms of "g's". If an aircraft undergoes a load factor of 3, it is also said to be experiencing 3 g's.

10.2.2 Turn Radius and Load Factor

Given above, the turn radius for a level turn is

$$R_t = \frac{m}{L}\frac{U_\infty^2}{\sin\phi}$$

We wish to have load factor in the expression for R_t. We begin by substituting W/g for m in the above equation.

$$R_t = \frac{W}{L}\frac{U_\infty^2}{g\sin\phi} = \frac{U_\infty^2}{gn\sin\phi}$$

Above, we showed that

$$\cos\phi = \frac{1}{n}$$

Using the trig identity

$$\cos^2\phi + \sin^2\phi = 1$$

we can replace $\sin\phi$ with a term that only has n:

$$\frac{1}{n}^2 + \sin^2\phi = 1$$

$$\sin\phi = \sqrt{1 - \frac{1}{n^2}} = \frac{1}{n}\sqrt{n^2 - 1}$$

Substituting into the original equation for turn radius, the turn radius can be expressed with only U_∞ and n:

$$R_t = \frac{U_\infty^2}{g\sqrt{n^2 - 1}}$$

From the above equation, it can be seen that the smallest turn radius can be achieved by having the lowest velocity and the highest load factor. This relation makes the importance of load factor evident.

Example: Sideward Forces

Traveling at 250 m/s, an aircraft performs a level turn, with the radius limited by the need to keep the g-level (load factor) below 7.0. What is turn radius R_t, and what is the roll angle ϕ?

Solution:
We begin by with the two equations of motion:

$$L\sin\phi = \frac{mu_\infty^2}{R_t}$$

$$L\cos\phi = W$$

The second equation can be used to find the roll angle.

$$\cos\phi = \frac{W}{L} = \frac{1}{\frac{L}{W}} = \frac{1}{n}$$

$$\phi = \cos^{-1}\frac{1}{n}$$

$$\phi = \cos^{-1}\frac{1}{7} = 81.8°$$

Dividing the centripetal equation of motion with the vertical equation of equilibrium yields:

$$\tan\phi = \frac{mu_\infty^2}{WR_t} = \frac{U_\infty^2}{gR_t}$$

$$R_t = \frac{U_\infty^2}{g\tan\phi} = \frac{(250)^2}{(9.8)(\tan 81.8°)} = 920.5m$$

The roll angle is 81.8° and the turn radius is 920.5 m.

10.3   Example: More Thrust Needed to Turn Without Losing Altitude

A pilot flying at 200 m/s decides to turn her airplane around, and needs to do that in the smallest possible radius, without losing altitude or speed. The limiting value of Load Factor (n= Lift/ Weight) is 2.
a) Find the smallest turn radius possible for this turn, assuming that the turn remains coordinated (this is the only case that we have discussed).
b) The drag changes with lift according to the rule

$$D = 0.03W + KL^2 \tag{10.1}$$

The L/D ratio in level flight was 20. By what percentage should the thrust be increased in order to keep the speed and altitude constant through the turn, assuming that the thrust remains exactly opposed in direction to the drag?

Solution: For straight and level flight, the lift must be exactly equal to weight (load factor $= 1$). When an airplane executes a coordinated turn at constant speed and altitude, it must increase its lift (usually by pitching up to increase angle of attack and hence the lift coefficient) and also increasing thrust to compensate for the increase in drag. In part (a) we deal with the increase in lift.

The load factor is limited to 2. The airplane has rolled to an attitude where the vertical component of lift balances the weight to keep it from losing altitude, and the horizontal (sideways) component of the lift provides the centripetal force to keep pulling towards the center, and balance the inertia (which shows up as the 'centrifugal force') So, if the roll angle is θ, then

$$LCos\theta = W \tag{10.2}$$

$$n = L/W \tag{10.3}$$

$$Cos\theta = 0.5 \tag{10.4}$$

$$Sin\theta = \sqrt{1 - (1/2)^2} = 0.866 \tag{10.5}$$

$$LSin\theta = MU^2/R \tag{10.6}$$

$$M = W/g \tag{10.7}$$

$$R = \frac{U^2}{ngSin\theta} \tag{10.8}$$

$$R = \frac{200^2}{(19.6)(0.866)} \tag{10.9}$$

$$R = 2357m \tag{10.10}$$

In Part (b) we deal with the increase in thrust required to keep the aircraft at constant altitude and speed. We assume that the thrust is used only to balance the drag and keep the aircraft moving at constant speed. (In most cases, when an airplane executes a tight turn, it must go to such a high angle of attack that the thrust vector points somewhat upward, and so part of the upward force balancing the weight also comes from the thrust. We are neglecting that here. So we assume that the thrust is equal to the drag.

The variation of drag with lift is given by

$$D = 0.03W + KL^2 \tag{10.11}$$

The weight of the aircraft remains constant, and K is a constant. Before the turn begins, the L/D is given as 20, and we know $L = W$.

So

$$0.05W = 0.03W + KW^2 \tag{10.12}$$

$$KW^2 - 0.02W = 0 \tag{10.13}$$

$$W(KW - 0.02) = 0 \tag{10.14}$$

$$K = 0.02/W \tag{10.15}$$

During the turn, L=2W

$$0.03W + 0.02(4W) \tag{10.16}$$

$$D = 0.011W \tag{10.17}$$

The change in thrust is

$$(\frac{0.11 - 0.05}{0.05})100 = 120\% \tag{10.18}$$

This makes more sense, because the lift doubled (increased by 100%) so the drag must have increased a great deal too, since some part of the drag is proportional to the square of lift.

10.4   Example on Descent with Velocity Vectors

An aircraft is flying straight and level at 100 m/s, when the pilot decides to start descending at a constant rate of 500 feet per minute without losing forward (meaning horizontal) speed. a) What is the Lift/weight and Lift/Drag ratio needed to do this? What is the glide path angle? b) If the airplane was headed along a vector $3i + 5j + 0k$, before starting the glide, what is the new flight direction vector? In our coordinate system, z is upwards from the Earth's surface, and the other directions are consistent with the 'right-hand-rule' convention for a Cartesian coordinate system where the axes are perpendicular to each other.

Solution: The descent rate is 500 feet per minute, which is 2.54 m/s. If θ is the glide path angle, then

$$Tan\theta = 2.54/100 \tag{10.19}$$

$$\theta = Tan_{-1}(0.0254) \tag{10.20}$$

This is 1.455 degrees, if you convert the answer in radians obtained above, to degrees, knowing that π radians is 180 degrees, where π is the famous constant that goes 3.14......, being the circumference of a circle of diameter 1. The magnitude of the direction vector is 5.83 (try squaring 3, 5 and 0, adding them and finding the square root). This corresponds to the 100 m/s speed. So, expressed in m/s, the velocity vector is initially

$$U = \frac{100}{5.83}(3i + 5j + 0k) \tag{10.21}$$

where i, j, and k are the unit vectors along the x, y and z directions respectively. When the downward velocity (component along the -z axis) is added, the velocity vector becomes

$$U = 51.46i = 85.76j + 2.54k \tag{10.22}$$

10.5   Range and Endurance: Breguet Range Equation

The range of an aircraft is the total distance it can travel on one load of fuel. The Breguet range equation is a good way to quickly estimate the range of an aircraft using some performance specifications of the aircraft. This equation assumes that the flight is steady and level. The derivation begins by considering the definition of velocity:

$$U_\infty = \frac{ds}{dt}$$

Because we are solving for range, or total displacement, we arrange this so as to have ds alone on the left side:

$$ds = U_\infty dt$$

Using the definition of thrust specific fuel consumption for a jet-propelled airplane, dt in the above equation can be replaced with some of the aircraft's performance parameters. Remember from our Propulsion chapter that the specific fuel consumption is expressed in units of (per hour) not (per second). We have to keep track of the factor of 3600.

$$sfc = -\frac{3600\dot{W}_{fuel}}{T} = -\frac{3600dW_{fuel}}{Tdt}$$

The negative sign is there because c_t is a positive value, but the weight of fuel is decreasing so the derivative is negative. Rearranging and solving for dt:

$$dt = -\frac{3600dW_{fuel}}{sfcT}$$

Substituting this into the displacement equation above:

$$ds = -\frac{U_\infty}{sfc}\frac{3600dW_{fuel}}{T}$$

We assumed steady, level flight. Thus, $T = D$ and we can replace T with D. We can also multiply the right hand side by L/D, as $L = W$; this is the same as multiplying by one. Also, it is reasonable to assume that $dW = dW_{fuel}$, or that the change in total aircraft weight is the same as the change in fuel weight. Applying these changes results in:

$$ds = -\frac{U_\infty}{sfc}\frac{L}{D}\frac{3600dW}{W}$$

Range is just the integration of ds. Thus we get:

$$R = \int_0^R ds = -\int_{W_0}^{W_1}\frac{U_\infty}{sfc}\frac{L}{D}\frac{3600dW}{W}$$

W_0 is the initial weight of the aircraft, when it is fully loaded with fuel, and W_1 is the final weight of the aircraft, when all the fuel has been used. Assuming that U_∞, c_t, and L/D are all constants:

$$R = \frac{3600U_\infty}{sfc}\frac{L}{D}\ln\frac{W_0}{W_1}$$

The full equation with the integral and all the quantities varying inside the integral, is known as the Breguet range equation. This equation shows three independent methods to increase range. Firstly, specific fuel consumption (c_t) should be minimized. Also, the fraction $U_\infty(L/D)$ must be maximized. The speed at which a plane flies affects its lift and therefore its L/D ratio, which is why we join U_∞ and L/D in one term; they are not independent of each other. Lastly, the ratio of the initial weight to the final weight (W_0/W_1) should be maximized. Fuel should compose as much of the initial aircraft weight as possible.

10.5.1   Fuel Consumed During Climb

For the climb phase, we will assume the fuel consumption for cruise plus an increment depending on the cruise altitude. The table below is constructed from data on large commercial airliners (Shevell).

Altitude	% of takeoff weight as added fuel consumption
20,000 feet	0.75%
30,000 feet	1.25%
35,000 feet	1.60%

In other words, if you take off at seal evel and level off at 20,000 feet, you will have used up 0.75% of the TOW in just the altitude change, besides the fuel consumption if the same forward distance had been covered at 20,000 feet in steady level flight. If you level off at 35,000 feet, the amount used up would have been 1.25%. These are of course ball-park estimates, because the actual result varies with the rate of climb used, the engine's thrust-specific fuel consumption, the use of flaps, the local weather, the climb path and any turns involved. For instance, taking off from Long Beach Airport, California is 'interesting' because noise regulations over the expensive real estate in the surroundings, forced a steep turn along with a rapid climb. On the other hand, if you take off from a high-altitude airport, this is one of the very few advantages of doing so, other than the spectacular scenery of mountainsides whizzing by.

10.5.2 Endurance

While range is the distance an aircraft can travel on one load of fuel, endurance is the amount of time the aircraft can stay in the air on one load of fuel. It is a measure of time, and as such is measured in seconds, minutes, or hours. There are many applications in which endurance might be more important than range. Observation or patrol aircraft are prime examples. They might need to watch a small area for large amounts of time, and not necessarily cover a lot of ground.

The derivation of the endurance for jet-propelled aircraft is similar to the derivation of the Breguet range equation. Again we assume steady, level flight with constant specific fuel consumption and a constant L/D ratio. The derivation begins with the definition of thrust specific fuel consumption:

$$sfc = -\frac{3600 dW_{fuel}}{T dt}$$

This time we are interested in solving for total time, so we isolate dt on the left-hand side:

$$dt = -\frac{3600 dW_{fuel}}{T(sfc)}$$

As we did in the Breguet range derivation, we will replace thrust with drag, multiply by L/W, and replace dW_{fuel} with dW.

$$dt = -\frac{1}{c_t}\frac{L}{D}\frac{dW_{fuel}}{W}$$

Endurance can be found by integrating from $t = 0$, when the aircraft is fully loaded ($W = W_0$), to $t = E$, when the aircraft has no fuel ($W = W_1$):

$$E = \frac{1}{c_t}\frac{3600 L}{D}\ln\frac{W_0}{W_1}$$

Similar to range, endurance can be increased by reducing the thrust specific fuel consumption sfc and by increasing the ratio of initial weight to final weight, W_0/W_1. Unlike range, endurance does not benefit from maximizing the product $U_\infty(L/D)$, as no U_∞ term appears in the endurance equation. Instead, endurance relies directly on the lift-to-drag ratio, L/D. Increasing the L/D ratio increases the endurance. This means that the speed for best endurance is the speed for minimum drag.

10.5.3 Minimum Velocity: Stall Speed

During steady and level flight, the lift of the airplane must equal the weight. These are both equal to the previously shown equation for lift:

$$W = L = \frac{1}{2}\rho U_\infty^2 S C_L$$

The right hand side of the above equation must equal the weight of the aircraft. The planform area S will probably remain constant (ignoring the possibility of some variable geometry wing), and so will the density, assuming level flight. To maintain a constant lift to support a (mostly) constant weight, U_∞ can decrease only if C_L increases. This is even more apparent if we rearrange:

$$U_\infty = \sqrt{\frac{2W}{\rho S C_L}}$$

As C_L increases, the speed required to stay in the air decreases. Thus, it is reasonable to say that the minimum speed at a given density and weight is:

$$U_{\infty min} = \sqrt{\frac{2W}{\rho S C_{Lmax}}}$$

10.6 Takeoff and Landing

10.6.1 Takeoff and Landing Distances

Net thrust is the thrust minus the ground roll friction, drag, etc. Assume net thrust is 0.2 W_{to}. Then $T - D = 0.2W_{to}$.

Kinetic Energy $= \frac{1}{2}mU^2 = \frac{1}{2}(\frac{W}{g})U^2 = 0.2WR$ where R is the takeoff run distance.

In distance R, we gained enough kinetic energy to be at takeoff speed, accelerating at the rate corresponding to net thrust of 0.2 W_{to}. Thus,

$$R = \frac{U^2}{0.4g}$$

The runway length should be twice this distance, in order to stop if the decision to abort takeoff is made at the takeoff speed.

10.6.2 Landing Procedure: Airliner

Descent to 5000 feet. Vectored to 12 miles downwind, make a 180-degree turn. Extend flaps and landing gear, and reduce speed to 150 mph. This leaves 5 minutes of final approach to do the flap deflections, landing gear deployment, lining up of the runway, etc.

10.6.3 Landing Procedure: Small Plane

Maintain constant angle of descent along a specified flight path (staying along the middle of the glide slope). Align with the runway centerline.

Arriving over the runway end, reduce power and increase angle of attack to level off with the landing gear a few feet off the ground. Use ground effect to continue flying level, as the speed reduces, and the aircraft slowly sinks to the ground.

Just before touchdown, flare (increase angle of attack, nearly to stall), and hold altitude. Rate of sink comes to very near zero, and as the speed comes below the stalling speed in ground effect, the main landing gear wheels touch the ground. Continue flying and bring nose slowly down until nose gear touches.

Use reverse thrust if any is available to slow down, otherwise roll with light braking, slow down to the speed for turning off the runway, and take the next taxiway turnoff as instructed by the tower.

10.6.4 Cross-Winds

Factors like cross-winds make life much more interesting. In the event of cross-winds, the airplane may be piloted down with the "controls crossed" (roll so that the lift vector is pointed into the wind, but use rudder in the opposite direction to keep from changing flight direction into the wind. As the airplane comes close to the runway, the wind will generally change quickly in speed and direction- and the pilot must compensate to keep the aircraft aligned with the runway. To land in a cross-wind, the airplane is held in a banked attitude, so that the landing gear on the side facing into the wind touches down first, then the other side, and the pilot holds the aircraft from swerving on the runway due to the greatly increased drag on one side.

Add gusts to the cross-wind, and the need for piloting skills becomes much more evident. Sometimes, to increase the rate of descent down to the runway, the pilot may fly with the aircraft actually sideslipping, alternately one way and then the other, and again controls crossed to induce a side-slipping motion. The reasoning here is that the drag is much greater in sideslip, so the forward speed is reduced and the angle of attack made a lot steeper.

10.7 Flight Envelope

"Flight envelope" is a term that refers to the performance range in which an aircraft can fly. If an aircraft stays within this envelope, the aircraft will be fine; if it ventures outside the flight envelope, catastrophic things can occur, as the aircraft is not meant to handle such things. An example flight envelope is the angle of attack range a wing can fly at. If a wing exceeds the angle of attack at which it stalls, flow separation will occur, lift will diminish, and the aircraft attached to the wing will be in trouble. However, if the wing stays within the envelope of acceptable angles of attack, no such misfortune will befall the aircraft.

10.7.1   Example

Gross weight = 200,000 N.
At altitude H, engine thrust is constant with speed, at 5000 N.
W/S = 8,000 N/m^2
$C_{Lmax} = 1.4$

Lowest speed (stall): $L = W = qSC_{Lmax}$

$$U_{stall} = \sqrt{\frac{2W}{\rho SC_{Lmax}}}$$

Lowest speed (thrust available): $T_a = D = qSC_D$

$$C_D = C_{D0} + \frac{C_L^2}{\pi (AR)e}$$

Highest speed (thrust available): $T_a = D = qSC_D$

10.8   Going Beyond

What speed should the pilot select? The answer depends on the mission. Back in the Wing Aerodynamics chapter we saw how there is a Speed for Minimum Drag. It is the speed where the lift-induced drag becomes equal to the lift-independent drag. Since the lift is equal to the weight for straight and level flight, this speed is also the speed for best L/D (lift-to-drag ratio). However this is usually too slow for most missions, unless one just wants to hang around. So the speed for maximum endurance (long time aloft without sinking) must be at least close to this speed. Note that this speed is not a constant: as the fuel is used up the weight and therefore the lift needed, decrease. So what happens? The wing loading also decreases with time. Does the speed for minimum drag increase or decrease? Think through that..

The speed for maximum range is different. In addition to staying up there, you must also cover distance, which means moving faster. So the speed for maximum range will be significantly higher than the speed for minimum drag (maximum endurance).

How fast can you fly? The ultimate limit for steady level flight is where drag equals the maximum thrust that you can generate, at any given altitude. That depends on your engines as well: how rapidly does thrust decrease as altitude increases?

But you CAN fly faster - in a dive. Much faster! So fast that the wings may break off. So there is a speed called V_{NE} The speed Never to be Exceeded. OK, the wings may not immediately break off, the engineers and builders were probably very good, but you don't want to find out.

If there are gusts, you can experience severe Load Factors due to that. So there is a lower speed which you must not exceed in case you encounter a gust. It is called the Maneuvering Speed. This is usually the limiting speed for most conditions. Unless one is extremely smart or foolhardy, it is best to stay below any of the Do Not Exceed speeds. One does not want to find out what happens if there is a gust, for instance. There is an old saying in the Pilot community:

There are Old Pilots and there are Bold Pilots, but there are few Old Bold Pilots.

10.9 Design Update

No doubt, this section has been quite long, and if you are like me you feel dazed by all the different rates and speeds and climbs and descents. But for now, we have no need to worry about most of those. Our main thrust is to see if our airliner design is basically feasible. All the rest can wait for detailed design analyses.

Let us recap

1. We started with a Requirements Definition.

2. We calculated the TOW using a Payload Fraction estimate.

3. Benchmarking also gave us all the other fractions as estimates.

4. We used benchmarking to pick a value for W/S, and from that we found S, aspect ratio AR, and the drag polar.

5. We estimated the thrust at sea-level takeoff with enough allowance to survive one engine failing.

6. We selected the engines, and got their thrust as a function of altitude and speed, and thrust-specific fuel consumption (which also depends on altitude and speed, but not as drastically as the thrust varies).

7. We verified that there was enough thrust at the selected cruise condition to overcome the drag. In other words, that the propulsion system specification was not controlled by the cruise requirement, but by the takeoff requirement. In reality, we should strive for a fine balance of these requirements. We will not get any brownie points for having engines that are larger than needed.

8. We also know the weight of the engines, and verified that we are well within the engine weight fraction allotted.

9. Now we use the Breguet Range Equation, integrating the fuel needed, from takeoff to landing. This gives the total fuel needed. In our case, we assume enough things like L/D to be constant so that the integration is a trivial exercise. Otherwise we could always step through in steps of time, which is what the integration does.

10. At this point, we know how much fuel is needed for the mission. We add that weight fraction to the payload and engine fractions. If we do not have 28 (supersonic) or 25 (subsonic) percent of the TOW left to build the structure, well, we cannot carry that much payload fraction. We must reduce the payload fraction, so that the TOW is higher. Of course then the engines must be bigger, and the fuel weight will be higher as well. This is the iteration needed.

11. On the other hand, if we find that we have too much weight left, more than what is needed to build the structure, then we should INCREASE the payload fraction. Iterate until things are finely balanced.

12. We now know that the design *closes*.

There are many other things to check using our Performance knowledge. How fast can the aircraft climb? What is the Steady Flight Envelope? We saw briefly what is meant by Flight Envelope. It is easy to calculate the boundaries of the flight envelope using a spreadsheet. The lowest altitude is of course the ground. Once in flight, even at ground level, there is a minimum speed: that at which you can stay up off the ground. The lowest speed is decided by the highest lift coefficient achievable before stall (power-off stall) or the power-on stall speed which is significantly lower, if we wish to live dangerously. The power-off stall speed is important: the ideal Kiss-Smooth Landing is when at landing we actually stall just as the main landing gear touches the ground. As altitude increases, the stall speed rises.

There is another limit on the lowest speed: the induced drag coefficient increases as the square of the lift coefficient. The drag could become so high that you simply do not have enough thrust to overcome it. This is rarely the limiter of speed at low altitude, but it may become the actual limiter at high altitude, so you must check this.

At high speed, the drag becomes too high for the thrust to counter it. For supersonic aircraft this limit behaves in a complex manner. Some jet engines rise in thrust as Mach number increases beyond 1. But beyond some Mach number, the thrust of the jet engine very rapidly drops to zero.

Knowing the lift and drag behavior and the thrust, you can calculate the takeoff field length needed. We assume that the ground friction may be as much as 10 percent of gross weight. Beyond takeoff, the craft must climb out quickly over any nearby obstacles. Similarly, an aircraft with plenty of thrust will be able to come in at a low speed and fairly high angle of attack (lift coefficient) and hence land on a shorter field.

We are about done with aircraft design for now. There is the issue of ensuring stability and controllability, but that is a much more complex issue, beyond ensuring basic static stability.

But.. did you notice that we STILL have not decided the configuration or shape of our aircraft? That comes after thinking about stability and control.

Summarizing, we list Design Step 5 as follows:

18. Find the fuel fraction needed for full mission range.

19. Determine propulsion and structure mass fractions.

20. Verify fuel weight and volume achievable-if not, iterate from # 14.

21. Develop steady flight envelop and plot. Determine extremes.

22. Verify ceiling, max speed and min speed.

23. Takeoff and landing field length; control surfaces/flaps needed.

10.10   Example Problems

10.10.1   Example 4: Fighter Aircraft Performance

The mission specification of a new fighter aircraft includes takeoff, climb to 60000 ft , cruise at Mach 2.0 for 1000 miles, loiter for 1 hour, engage in close combat at Mach 0.7 for 30 seconds, fire off 6 missiles, and return home with a supersonic dash at Mach 3.0 at 60,000 feet for 100 miles, followed by Mach 0.7 cruise at 40,000 feet and a vertical landing on an unprepared field. It is found that the payload (armaments) weighs 10,000 lbs, and that the payload fraction of Maximum Takeoff Weight is 0.2. Find the Max. T.O.W.

At 80 percent of the Max T.O.W., at sea-level, the thrust-to-weight ratio has to be at least 1.4 (the F-15s can accelerate going straight up; the F-22 thrust to weight ratio is given somewhere as 1: 0.72, or 1.39). The best engine suitable for use on this fighter has a maximum sea-level static thrust of 38,000 lbs. How many engines do you need on this aircraft? What is the takeoff thrust? What is the maximum acceleration assuming that the weight is at Max TOW at takeoff, if you take off and rotate so that the fighter is pointing straight up? If you divide this by $9.8m/s^2$, you can express the acceleration in "G"s. How many "G"s does the pilot experience during this phase? (don't forget the 1G that she experiences all the time due to Earth).

Solution

Let us list the given information, and convert it to Standard International (SI) units of meters, kilograms, seconds, with forces in Newtons. I use a Microsoft Excel (or Apple Numbers) spreadsheet to do these calculations. You can enter a formula in each cell and copy that to other

Parameter	Value in given units	SI or dimension-less
Altitude ceiling , feet	60000	18288
Cruise Mach Number		2
Supersonic cruise range, miles	1000	1600000
Return cruise Mach		0.7
Subsonic cruise altitude, feet	40000	12192
Vertical landing, unprep field		
Payload, lbs	10000	44545
Payload fraction at TOW		0.2
T/W, sea level at 80 percent of TOW		1.4
Single engine sl thrust, lbs	38000	169273

Table 10.1: Mission Specification

Max TOW is the payload divided by payload fraction	222727
80 percent of Max TOW	178182
Thrust needed to get the specified T/W	249455
# of engines needed, given single-engine thrust	1.47
Round-off to whole number of engines	2
Max thrust at sea level (sl) takeoff	338545
Vertical acceleration	5
Vertical acceleration, Gs	0.52
What the pilot feels in Gs, including the 1G felt in level flight	1.52

cells to repeat the calculation. For instance, in my spreadsheet, the weapon payload in pounds (lbs) is given in Cell C8, as 10,000. To convert this to Newtons in Cell D8, I use the formula: $= C8/2.2 * 9.8$. People use the term 'lbs' denoting 'pounds' to mean weight. One kilogram of mass has a weight of 2.2 lbs, or 9.8 Newtons (N). That calculates the result for payload as 44545 N. You can guess the formulae used for the other cells.

Mission specification

Derived quantities

The climb rate is an important specification for air superiority fighters and light interceptors. In air combat, height is a strong advantage, and she who gets to high altitude faster therefore has an advantage. The maximum climb rate exceeds 60,000 feet per minute but cannot be maintained all the way from takeoff, so it takes perhaps 1 to 2 minutes if one is seriously interested.

10.10.2   Example 5: Turn Radius

A pilot flying at 200 m/s decides to turn her airplane around, and needs to do that in the smallest possible radius, without losing altitude or speed. The limiting value of Load Factor (n= Lift/

Weight) is 2. a) Find the smallest turn radius possible for this turn, assuming that the turn remains coordinated (this is the only case that we have discussed in class). b) The drag changes with lift according to the rule

$$D = 0.03W + KL^2 \tag{10.23}$$

The L/D ratio in level flight was 20. By what percentage should the thrust be increased in order to keep the speed and altitude constant through the turn, assuming that the thrust remains exactly opposed in direction to the drag?

Solution: For straight and level flight, the lift must be exactly equal to weight (load factor = 1). When an airplane executes a coordinated turn at constant speed and altitude, it must increase its lift (usually by pitching up to increase angle of attack and hence the lift coefficient) and also increasing thrust to compensate for the increase in drag. In part (a) we deal with the increase in lift.

The load factor is limited to 2. The airplane has rolled to an attitude where the vertical component of lift balances the weight to keep it from losing altitude, and the horizontal (sideways) component of the lift provides the centripetal force to keep pulling towards the center, and balance the inertia (which shows up as the 'centrifugal force') So, if the roll angle is θ, then

$$LCos\theta = W \tag{10.24}$$

$$n = L/W \tag{10.25}$$

$$Cos\theta = 0.5 \tag{10.26}$$

$$Sin\theta = \sqrt{1 - (1/2)^2} = 0.866 \tag{10.27}$$

$$LSin\theta = MU^2/R \tag{10.28}$$

$$M = W/g \tag{10.29}$$

$$R = \frac{U^2}{ngSin\theta} \tag{10.30}$$

$$R = \frac{200^2}{(19.6)(0.866)} \tag{10.31}$$

$$R = 2357m \tag{10.32}$$

In Part (b) we deal with the increase in thrust required to keep the aircraft at constant altitude and speed. We assume that the thrust is used only to balance the drag and keep the aircraft moving at constant speed. (In most cases, when an airplane executes a tight turn, it must go to such a high angle of attack that the thrust vector points somewhat upward, and so part of the upward force balancing the weight also comes from the thrust. We are neglecting that here. So we assume that the thrust is equal to the drag.

The variation of drag with lift is given by

$$D = 0.03W + KL^2 \tag{10.33}$$

The weight of the aircraft remains constant, and K is a constant. Before the turn begins, the L/D is given as 20, and we know $L = W$.

So

$$0.05W = 0.03W + KW^2 \tag{10.34}$$

$$KW^2 - 0.02W = 0 \tag{10.35}$$

$$W(KW - 0.02) = 0 \tag{10.36}$$

$$K = 0.02/W \tag{10.37}$$

During the turn, L=2W

$$0.03W + 0.02(4W) \tag{10.38}$$

$$D = 0.011W \tag{10.39}$$

The change in thrust is

$$(\frac{0.11 - 0.05}{0.05})100 = 120\% \tag{10.40}$$

This makes more sense, because the lift doubled (increased by 100%) so the drag must have increased a great deal too, since some part of the drag is proportional to the square of lift.

10.10.3   Example: Aircraft Performance

NOTE: For the following problems, please feel free to solve simultaneous algebraic equations, do some derivation, etc. as needed.

1. An aircraft is climbing. The flight path is inclined at 3 degrees to the horizontal. The weight of the aircraft is 2 million Newtons. The airplane is in equilibrium flight. The thrust is 200,000 Newtons. Find the lift and drag.

2. An aircraft is flying straight and level, on autopilot. The lift-to-drag ratio is 18. Suddenly the pilot sees his buddy Bob flying his plane parallel, a few miles away, and decides to race him. He increases the thrust, so that the aircraft accelerates. At the new speed, the thrust is 1.4 times the drag force. The L/D ratio is still 18. Find the values of the acceleration along each component in g's. Now find the magnitude and direction of the resultant acceleration vector.

3. An aircraft makes a coordinated turn at constant speed, without losing height. The turn radius is 600m. The aircraft speed is 290 m/s. Find the "g"s acting on the pilot (in addition to the 1-g acting downwards).

4. An aircraft banks into a coordinated turn. The lift is twice the weight of the aircraft. The speed of the aircraft is 200 m/s. Find the radius of the turn so that the radial acceleration is kept to 9g's.

5. Traveling at 250 m/s, an aircraft performs a coordinated turn, with the radius limited by the need to keep the g-level (load factor) below 9.0. What is the radius in meters, and what is the bank angle?

6. A fighter aircraft does a coordinated turn, while flying at 200 m/s. The turn is limited by the pilot's capability to take 'Gs', in this case limited to 7. Find the turn radius.

7. An aircraft flying at 250 m/s is in a coordinated turn, where it banks 30 degrees to the right, while maintaining altitude and speed. Find the radius of the turn if the load factor is to be kept below 7.

8. An airline pilot flying at 300 m/s executes a coordinated turn without losing altitude or speed. The limiting value of Load Factor is 5. Find the turn radius.

9. Traveling at 250 m/s, an aircraft performs a coordinated turn, with the radius limited by the need to keep the g-level (load factor) below 9.0. What is the radius in meters, and what is the bank angle?

10. While conducting an incredibly precise flight test, a fighter pilot who was flying straight and level at 21 deg. angle of attack at Mach 0.3 at 10,000 feet ISA , pulls up to 25 deg. angle of attack, while simultaneously adjusting his thrust to compensate for the drag increase. The wing loading in straight and level flight is 80 psf . This aircraft has a delta wing whose lift-curve slope is approximately constant at 0.2π per radian for $20 < \alpha < 26$ degrees (note:

this is MUCH lower than that for an ideal airfoil section, see below!) Calculate the upward acceleration of the aircraft in G's, assuming that the thrust vector stays aligned with the horizontal (i.e., do not worry about the thrust).

11. A pilot has to execute a coordinated 90 - degree turn through a mountain canyon, where she cannot afford to change altitude or speed, or tolerate side-slip (which is what happens if the turn is not coordinated, and would result in hitting canyon walls), and the turn has to be made with a radius of curvature not to exceed 800meters. Since there is a Klingon vehicle pursuing her, she must do this at the highest speed possible. The load factor limit means that her vehicle must not exceed $80 m/s^2$. How fast can she be flying as she executes the turn? What is the roll angle in degrees?

12. As she straightens out from the turn, the pilot of the problem above rolls the airplane back to a level attitude at the same speed, but holds the angle of attack, in order to rise rapidly. The airplane drag varies proportional to dynamic pressure, and the lift-to-drag ratio in straight and level flight before starting the turn was 12. What is the percentage increase in thrust needed, over the straight-and-level condition?

13. What is the upward acceleration of the aircraft in the above problem?

14. The flight path vector of a missile is along $5x + 3y + 8z$. It's speed is 4000 m/s relative to the ground. It fires a warhead, giving it an additional velocity component of 200 m/s, entirely along the y direction. Find the flight speed of the warhead and its flight path vector relative to the ground, all components expressed in units of m/s.

15. An aircraft is flying straight and level at 100 m/s, when the pilot decides to start descending at a constant rate of 500 feet per minute without losing forward speed. a) What is the Lift/weight ratio needed to do this? What is the glide path angle?

    b) If the airplane was headed along a vector $3x + 5y + 0z$, before starting the glide, what is the new flight direction vector? In our coordinate system, z is upwards from the Earth's surface, and the other directions are consistent with the 'right-hand-rule' convention for a Cartesian coordinate system where the axes are perpendicular to each other.

16. You have to come up with a concept for a supersonic aircraft to send a small force of Special Forces soldiers across the world. The passengers in this case have an average weight of 190 lbs, and there is 80lbs of equipment per soldier. The force has 100 soldiers (officers included). In addition, the aircraft must carry 5000 lbs of other equipment.

    Assume a payload fraction of 0.15. The range will be roughly 8000 miles. Estimate the aircraft TakeOff Weight. The wing loading will be 70 lbs per square foot. The wing span is limited to 200 feet. The wing planform is approximately a triangle. Find the wing area and the wing root chord.

17. At an altitude where density is 20% of its standard sea level value, An aircraft has 5% more thrust available than it has drag. The aircraft is flying at 200 m/s, with a CL/CD value of 17, when the pilot is asked to climb at the maximum climb rate possible, while keeping the flight speed at 200m/s. Find this climb rate, and determine whether the aircraft is already at its 'ceiling' as we defined it for this assignment.

10.10.4 Example: Glide Performance

An aircraft is flying straight and level at 100 m/s, when the pilot decides to start descending at a constant rate of 500 feet per minute without losing forward (meaning horizontal) speed. a) What is the Lift/weight and Lift/Drag ratio needed to do this? What is the glide path angle? b) If the airplane was headed along a vector 3x + 5 y + 0z, before starting the glide, what is the new flight direction vector? In our coordinate system, z is upwards from the Earth's surface, and the other directions are consistent with the 'right-hand-rule' convention for a Cartesian coordinate system where the axes are perpendicular to each other.

Solution: The descent rate is 500 feet per minute, which is 2.54 m/s. If θ is the glide path angle, then

$$Tan\theta = 2.54/100 \tag{10.41}$$

$$\theta = Tan_{-1}(0.0254) \tag{10.42}$$

This is 1.455 degrees, if you convert the answer in radians obtained above, to degrees, knowing that π radians is 180 degrees, where π is the famous constant that goes 3.14......, being the circumference of a circle of diameter 1.

The magnitude of the direction vector is 5.83 (try squaring 3, 5 and 0, adding them and finding the square root). This corresponds to the 100 m/s speed. So, expressed in m/s, the velocity vector is initially

$$U = \frac{100}{5.83}(3i + 5j + 0k) \tag{10.43}$$

where i, j, and k are the unit vectors along the x, y and z directions respectively. When the downward velocity (component along the -z axis) is added, the velocity vector becomes

$$U = 51.46i = 85.76j + 2.54k \tag{10.44}$$

10.10.5 Example: Range

A pilot embarks on a cruise segment, where he has to get to an airport 4000 kilometers away, starting at a speed of 250m/s, an altitude where density is 25% of sea level value . The thrust available for the aircraft varies with altitude, inversely proportional to density. The aircraft gross weight at takeoff from sea level was 1.5 million Newtons. At the cruise altitude it has a tsfc of 0.4. At the start of the cruise, the wing loading is 8000 $\frac{N}{m^2}$, and 25% of the weight at that point is usable fuel. The pilot flies so that the variation of lift and drag coefficients through the cruise can be approximated on average, that drag coefficient is 0.03 and lift coefficient is 16 times that. Will the pilot make it to the destination airport? Why/Why not?

Chapter 11

Stability and Control

11.1   Introduction

When you launch a paper airplane, the most frequent cause of an unsuccessful flight is that the airplane flips out of control. This is usually because the airplane is not statically stable. The flight starts out promising enough, but then the nose keeps pointing further up, until the angle of attack becomes large. The drag becomes high. The speed drops. And then the airplane drops. By carefully adjusting the weight distribution with respect to its lift distribution, or deflecting some control surfaces to change the lift forces, the flight characteristics can be greatly improved. In the case of the paper airplane, adding a pebble at the nose will probably make the vehicle fly much better. We will see why.

The nomenclature for the six 'degrees of freedom' of a flight vehicle is shown below. The six degrees of freedom are:

11.1.1   Translation

1. x: forward (or rearward) motion $\mathbf{U}$

2. y: sideslip $\mathbf{V}$

3. z: climb or descent (assuming airplane level): $\mathbf{W}$

11.1.2   Rotation

1. l: roll ϕ

2. m: pitch θ

3. n: yaw ϵ

The control surfaces may be placed in various ways, as seen in these pictures

F-15 with close-coupled canards and all-moving tails.

Figure. 11.1: Different types of control surfaces and placements. Above left: tiltrotor aircraft, with wing flaps deflected, Above right: Uninhabited fixed-wing aerial vehicle. Bottom: F-15 fighter aircraft in a modified version with canards. Credits: All images, United States Government.

The Longitudinal degrees of freedom are: **U**, **W**, θ The Lateral degrees of freedom are **V**, ϕ, ϵ

11.2   Important Points

Every part of the aircraft has mass, and therefore the force of gravity acts on every part. This is the weight. We can represent the total weight to be a force acting through a single point on the body: this is the **center of gravity**. If we hold the body at this point in still air, it has no tendency to fall over to any side. In other words, this is the point located such that the moment due to gravitational force on every part of the body, taken about this point, cancels out.

Similarly to the process of finding the center of gravity, we can define a **center of pressure**. We have defined lift as the force acting on a wing perpendicular to the freestream, and drag as the force acting parallel to the freestream, but have not really visualized how these forces act on a wing. There are pressure and shear stress distributions acting over the surface. It is possible to find the net effect of these distributions, and represent them with a single concentrated force **R**, acting at a specific point. The specific point at which this resultant force **R** acts is the center of pressure, denoted by c.p. The force **R** can then be broken down into components perpendicular and parallel to the freestream, or lift and drag. Thus the lift and drag have no moment about the center of pressure.

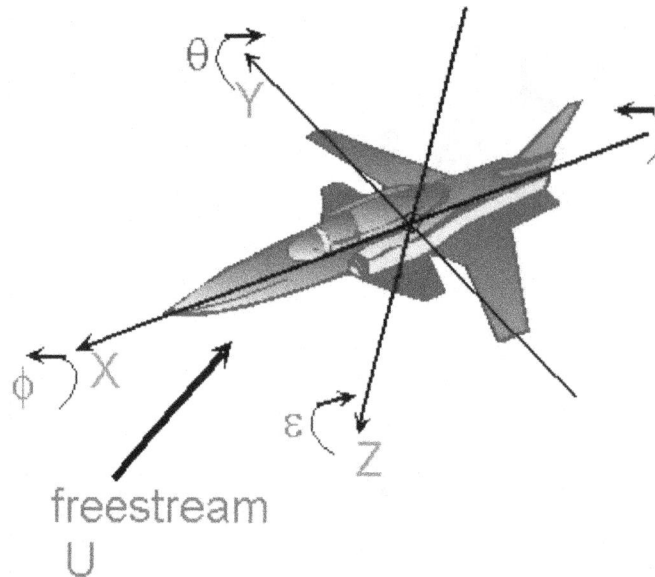

Figure. 11.2: The 6 degrees of freedom

The center of pressure of course, depends on the pressure distribution. The pressure distribution can and does change as flight conditions change. For instance, at low speeds and small angles of attack, the center of pressure of a symmetric airfoil section is located at the quarter-chord point. In other words, at 25% of the chord downstream of the leading edge. If the airfoil is cambered, the c.p. is located a bit further downstream.

As the speed increases beyond 30 percent of the speed of sound, the effects of density change due to change in speed, become significant. The center of pressure moves downstream. In supersonic flow, the center of pressure is usually seen at mid-chord; i.e., 50% of the chord downstream of the leading edge.

Aerodynamic Center The aerodynamic center of an airfoil is the point about which the pitching moment coefficient is independent of the lift coefficient. About the aerodynamic center, the pitching moment stays constant, within a reasonable range of orientations of the airfoil (angle of attack) with respect to the flow.

11.3   Static Stability

An aircraft is statically stable if it can recover from small disturbances by itself. It is statically unstable if the disturbance gets amplified and the aircraft does not recover.

Example of Static Instability: Imagine that the center of pressure on a wing is ahead of the center of gravity, as shown below. Suppose that the angle of attack α increases due to some small, instantaneous gust. The coefficient of lift, and therefore lift, would increase because of the increase in α. The change in pitching moment is nose-up, which further increases α. Thus, a slight increase in α would continue to get larger; this is statically unstable.

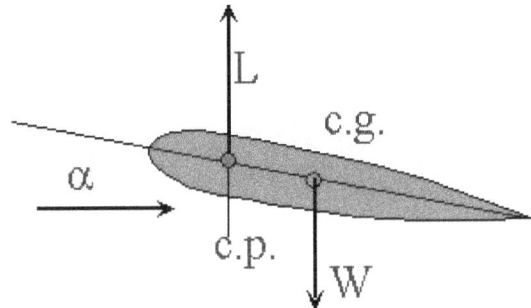

Figure. 11.3: Center of pressure CP in front of the center of gravity CG.

If the center of pressure were located behind the center of gravity, lift would still increase if angle of attack were increased. However, this time the lift force is acting behind the center of gravity (at the c.p.), and causes a nose-down pitching moment. Thus the system acts to reduce α if α is suddenly increased, so this system is statically stable. Likewise, as shown in Figure 11.5, a side force applied behind the center of gravity is stable, but one applied ahead of the center of gravity can be destabilizing.

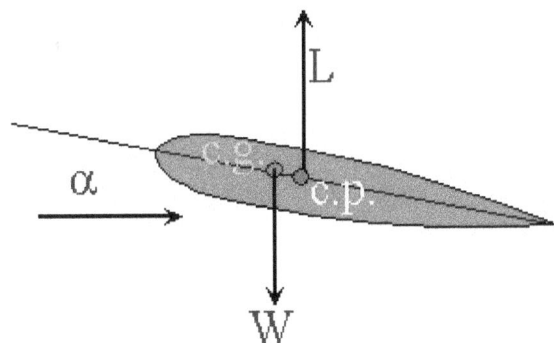

Figure. 11.4: CP behind CG.

Thus, **center of pressure behind center of gravity is stable.**

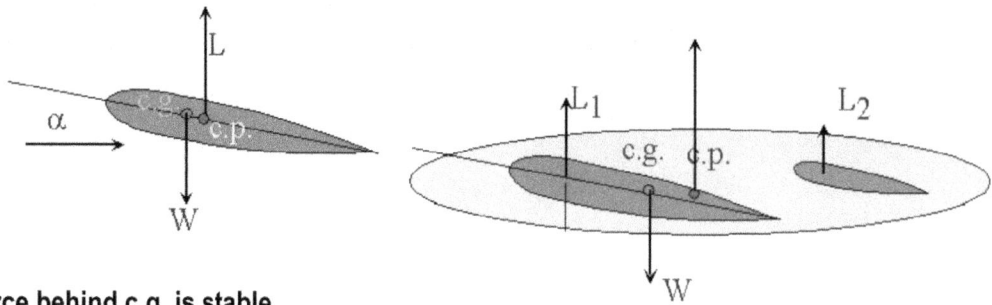

Side force behind c.g. is stable.

Side Force
Ahead of
c.g. is
destabilizing

Side Force
aft of c.g. is
destabilizing

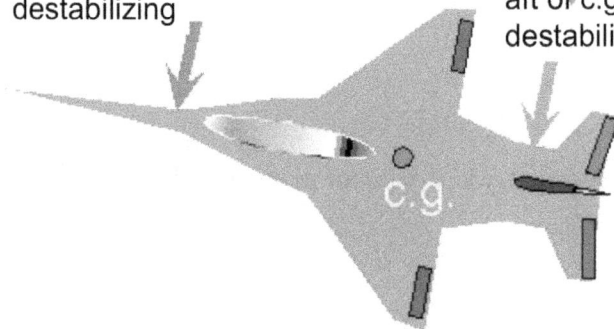

Figure. 11.5: Longitudinal and Lateral Stability

11.4   Aerodynamic Control Surfaces

We introduced roll, pitch, and yaw introduced as rotations about axes of the aircraft. In order to induce or reduce these rotations, aircraft need control surfaces. These were discussed in the chapter on motion and dynamics. Here we repeat these, now that we know aerodynamics, to see how they work. The pilot's control from the cockpit by tradition, is to tilt the control column (stick). Tilting it forward pushes the nose down by tilting the elevators down and thus increasing lift on the tails, In cases where a front-located set of canards is the main pitch control, read these with the appropriate modifications.

Elevator: Pitch Control:Elevators are flight control surfaces found on the horizontal part of the tail structure of most aircraft.

Deflecting the elevator up causes the aircraft nose to *pitch* up.

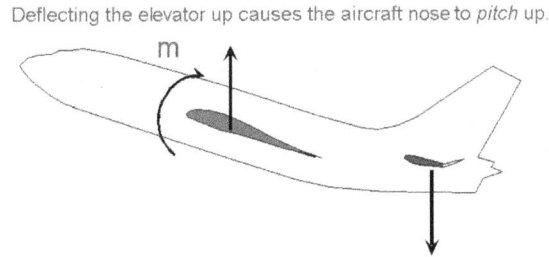

Figure. 11.6: Deflecting the elevator up causes the aircraft nose to pitch up.

Aileron: Roll Control Ailerons are control surfaces located at the trailing edge of aircraft wings. Deflecting an aileron is like cambering the airfoil section of the wing. It changes the lift at the same angle of attack. If the aileron is deflected down, the wing produces more lift; an aileron that is deflected up causes the wing to produce less lift (or even negative lift). Because of this force imbalance, the aircraft rotates about the roll axis.

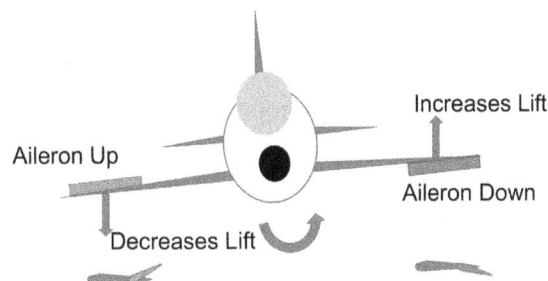

Figure. 11.7: Deflecting the ailerons causes the aircraft to roll.

Typically, ailerons are connected in such a manner that when one deflects up when the other deflects down. Tilting the control column to the left makes the airplane roll to the left.

Rudder: Yaw Control The rudder is the vertical control surface found on the tail structure of most aircraft. If the lift (side force in this case) on the vertical tail is changed, the aircraft tends to yaw. The aircraft must then roll to avoid sideslipping. In aircraft, yaw is not used to turn aircraft. As was shown in the performance chapter, aircraft roll in order to turn. The main use of the rudder is to make minor changes and corrections to keep the nose pointed into the turn as the aircraft goes through a turn. This avoids 'sideslip'. where the aircraft appears to be traveling sideways.

Deflecting the rudder to the right causes the aircraft nose to
yaw to the right.

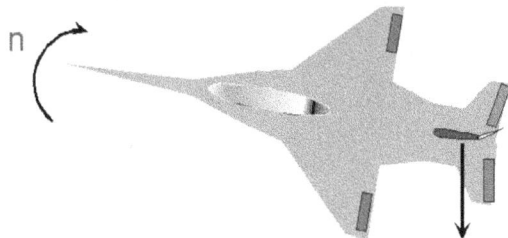

n

If the lift (side force) on the
vertical tail is changed, the
aircraft tends to yaw.
Then the aircaft must roll to
avoid sideslipping.

Figure. 11.8: Deflecting the rudder to the right causes the aircraft nose to yaw to the right.

Sideslip makes one look very clumsy as a pilot. The accomplished pilot manages to keep the nose pointed steadily in the direction in which the aircraft is traveling. She also pulls the control column back a bit to increase the aircraft's lift coefficient, because the roll requires part of the lift to be acting sideways. The aircraft would descend if the lift coefficient is not increased. When the lift coefficient is increases (or if the aircraft slips sideways!) the drag goes up, so the speed would go down, unless the thrust is increased. So the pilot also increases thrust during the turn, or takes this step precisely and deliberately just before the turn is commenced. By the time the propellers or jet engines get the message and the thrust actually increases, the aircraft is already turning, so it all appears totally smooth and coordinated.

Sideslip is not all bad. It is a great tool for a pilot to (what else?) increase drag and lose height. Also, to fight an inconvenient cross wind while trying to stay locked to a runway during landing approach. Pilot manuals suggest two types of maneuvers while approaching in a cross wind. One is to successively roll right and left, zigzagging a bit from the straight line approach. But if one is comfortable with the sideslip, a better way may be to come nearer the runway at a higher altitude, roll into the wind and but not coordinate the turn. Stay aligned with the runway, keeping the runway touchdown point locked in one's vision. To the outside observer the aircraft appears to roll, but come in straight along the descent path. Because of increased drag the airplane loses height more rapidly.

On my second attempt to fly solo, my dear instructor had breezily given approval for me to go ahead. That removed one of my main excuses to not fly. So I had to get in the aircraft and do the pre-flight checks. The day was windy, with gusts. I finished the checks. Another excuse gone. I asked the Tower for permission to taxi. No luck - permission granted with the usual complete lack of emotion or apparent sympathy. I taxied, the plane buffeting as the wind gusts shook it (and me inside). I reached the end of the taxiway and stopped. All that remained was a left turn onto the runway- and accelerate. I looked for any excuse. Yes! There was a plane coming in to land. I waited.

The plane was obviously flown by an expert (note that any pilot who had completed two solos would be more expert than I, by definition). He came in with strong sideslip. As he neared the treetop level the wind direction changed: he smartly rolled and sideslipped in the opposite direction. The wind gusted severely at low levels. I saw him rapidly roll and yaw from one side to the other as he passed me, almost snap-rolls like I had read about fighter pilots. He touched down with the nose pointed to one side, then allowed the friction to pull the craft level, and taxied smartly out of my range of vision.

I could feel the cold fear. If I took off (as I was sure I could, though with some adventure), how in the world would I come down and execute a landing like THAT? How long could I hang around waiting for the weather to clear? WHY did the Tower not say, *"Hey YOU! Didn't you tell me a few days ago that you were doing your first solo? It's way too windy and gusting. You ain't competent to fly today. Go back to ramp and go home!"*

But that's not how it works. I was The Pilot In Command: the life-or-death decision was entirely up to me. All excuses were now gone. The Tower expected me to start rolling in 2 seconds for takeoff. The old saying of the Pilot's Lore rang in my ear.

It is better to be down here wishing you were up there..
than to be up there wishing you were down here.

The decision was made, which is why I lived to tell this tale. Common sense over pride. Fear over stupidity. *"Golf Tango Niner Five Niner requesting permission to abort takeoff and return to ramp." "Golf Tango Niner Five Niner, permission granted. Taxi on to the runway and take taxiway Hotel."* I swear I could hear relief and joy in the controller's voice. Relief! That low-speed ride down the runway to the next exit was scary enough: the gusts buffeted me, but I was happy. Returned to the ramp, locked down the aircraft, and went home, shaken, wiser, but safe.

I am surely making a big deal out of nothing. If I had stepped on the gas and taken off, my speed would have made the gusts appear small and I would have swiftly climbed out of the low-level gusts. Returning for landing, my training and courage (yeah!) and trained skill (yeah, *right!*) would have stood me in good stead, I would have rapidly gained experience and confidence with each snap roll and "crab" (the Insider's Name for the sideslipping descent). After all, sideslip is one thing I could surely execute, having done that every time the instructor asked me to do a coordinated turn ("Right Rudder! Right Rudder!" his voice rings in my ear). I would have done a perfect touchdown, though a bit sideways, and immediately corrected to taxi smoothly and proudly.

Or maybe I would have crashed in a heap. Drifted off the runway at the last moment, and tried to overcorrect. One side touched down on grass, plane yanked violently to that side while rolling the other way. The laws of dynamics are a heck of a lot less sympathetic than that kind Controller watching over me. Not a chance worth taking. It was my job as Pilot In Command to do what was right for the Passengers (me) and for the aircraft and for the airfield and neighbors. And safe is always the right thing in our business. To quote that other saying:

There are Bold Pilots, and Old Pilots. But there are no Old Bold Pilots. Not in civilian airspace.

Please visit the website titled "Crosswind Landing Procedures" at [130] for an excellent discussion, and some terrific pictures. If you search under "Crosswind landing" on the Internet, you will find many videos, which I would not recommend watching if you have bought an air ticket and are planning to fly soon. Hong Kong's Kowloon Airport, now replaced by a newer airport away from the city, was the site for many of the most exciting videos on this subject, bringing Boeing 747 airliners in over crowded high-rise neighborhoods, sometimes appearing to fly between buildings, to land in fairly extreme cross-winds. One site called 'Top 20 Crosswind Landings" had over 13 million views by July 2018, and all of them appeared to have happy endings, though most of those were last-second aborts taking off again at full power. Pilot's decision.

11.4.1   Design Update

What remains in our Conceptual Design? Well... only to decide the configuration of the aircraft. We know there must be a fuselage with a certain volume, and wings with a certain planform area and aspect ratio. We know how many engines and of what type we need. Now it time to arrange these in the best manner. Stability is an important consideration here: We must ensure static stability, unless it is a modern fighter with Relaxed Static Stability or outright statically unstable and computer-controlled. For supersonic airplanes, this is more involved. The center of pressure moves back as the flight speed increases, and settles around mid-chord instead of the quarter chord at low speeds. The center of gravity must be moved accordingly. This is done by pumping fuel from forward tanks towards mid-wing and aft tanks. Then as the aircraft slows down, the remaining fuel must be pumped back to the front. This means that rapid acceleration and deceleration of high-speed aircraft is not a wise move.

The horizontal tails and vertical tails, the rudders and the ailerons, are all sized based on the requirement for stability and control.

11.4.2   Example Problems

An aircraft has a vertical tail of area 10 m^2, and the centroid of that area is located at a distance of 10m downstream of the center of gravity, and 10.5m downstream of the center of pressure. Is this aircraft stable (precise term: does it have positive longitudinal stability, but donÕt worry about that)? How much is the vertical tail volume?

Chapter 12

Structures

12.1 Introduction

Figure. 12.1: Image A91-0261-15 – Circa 1934 photo of Hangar 1 with the dirigible U.S.S. Macon. Credit NASA Ames Research Center archive image.

Figure 12.1 shows the blimp hangar at NASA Ames Research Center. It is an example of what can be done with truss structures. Many very large, ultra-light structures proposed for construction in Space are trusses. We will discuss them further below. Aerospace structures have to be extremely light and yet strong where needed. Mass must be used well, and exactly where it is needed. The need to keep the total mass as low as possible, reduces the available safety margins. Since safety margins are a measure of uncertainty, this means that in aerospace structures, it is essential to reduce uncertainty to the minimum. Detailed analysis of aerospace structures is today often done using the Finite Element Method, typically using large commercially-available computer programs with Graphical User Interfaces. This makes it ever more important that the engineer who uses these programs must have a very good 'feel' for what the results should be - and a suspicious mind that directs attention to any optimistic occurrences.

Legend holds that the first breakthrough success of the minicomputer called the Fourier Analyzer was when the inventors were allowed to instrument a tall building that was under construction, and use their new gizmo to analyze the vibratory modes of that structure. They found that the first model of 'freedom' was at zero frequency. Ah! Experimental Error? No, the inventors knew their stuff. They took the result to the owners, who were smart enough to institute enquiries. Someone had forgotten to put in some important bolts or pins or Crazy Glue, whatever they used to build skyscrapers. The skyscraper was not anchored or fixed to the foundations! Forget earthquakes, a good breeze might have blown it over. (Did I know that when the professor related that story in the Graduate Structural Dynamics class? Of course not. I asked someone who knew Structures: "psst! What's so special about Mode 0?" and they were kind enough to explain, and reserve their laughter at my ignorance for the Friday evening party).

To design structures we have to understand stresses. *Stress* is *force per unit area*. This has the same units as pressure, but is more complicated. Whereas the force due to pressure is always taken as acting normal to (perpendicular to) the surface on which the pressure acts, stresses are vectors. We can have shearing stresses (tangential to a solid surface) in addition to normal stresses.

12.2   Stress and Strain

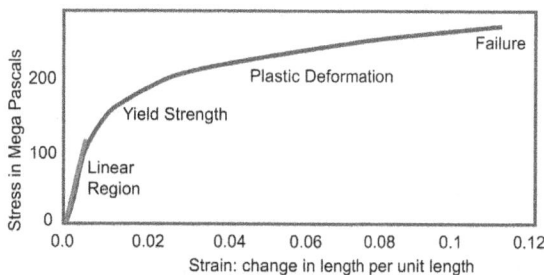

Figure. 12.2: Typical relation of tensile strain to stress, for an aluminum alloy. The slope of the linear region at the low end, is the Young's Modulus of Elasticity. Beyond the Yield value of stress, strain increases substantially until the eventual failure of the material.

Under stress, almost all materials change dimensions. This is called Strain. Strain is defined as the *change in dimension divided by the dimension*. Thus the tensile strain would be the change in length of a strucrural element, divided by its length. A stress-strain relation may be like that shown in Figure 12.2. Most materials used in building structures show very little strain (small percentages of their dimensions) as long as the stress is below the Yield Stress. Yield Strength is thus defined as the lowest stress that produces a permanent deformation in a material. Since the precise level is hard to identify, the Yield Strength for metals such as aluminum is the stress that causes a plastric strain of 0.2 percent. Materials used in aerospace structures, particularly metals, also exhibit the nice property that the strain is small, and is directly proportional to stress, upto pretty high levels of stress. This makes it easier to predict their behavior. The stress increase required to induce a given increase of strain stays approximately constant, until the Yield Stress is reached: beyond that, a very small increase in stress, or even no increase, may cause a rise in strain. Some materials may become plastic, experiencing very large amounts of strain without breaking. Others may break.

The slope of the above curve, in the linear range, is called the Modulus of Elasticity: the stress needed per unit strain. The higher the Modulus, the stiffer the material, but not necessarily stronger.

The strength is measured by the Yield Stress, to be safe. Actual failure should occur only at significantly higher levels of stress: the practical definition of a failure is that no further increase in stress is resisted. Either the material elongates so that it constricts at the point where stress is higheset and that increases the stress further increases since the area over which the force is divided, decreases. The constriction rapidly proceeds to where the material separates. Or a crack (a small separation) may start growing, which rapidly aggravates the situation. Designers strive to stay well away from the Yield Stress, building in a Margin Of Safety. That said, note that modern aerospace designers confidently use pre-stressed components that have already been driven beyond the yield point, in some cases where they carefully know what they are doing. Later we may see how Buckling is a feared type of failure for thin rods and beams; modern designers use *post-buckling* structural components which then offer much greater strength. But for now, avoid such things. Yield is bad; it usually means that when the load is removed, the component does not return to its original dimensions because its interior molecular arrangement has changed. It is no longer in a range of *elastic deflection*. Failure is not an option at all.

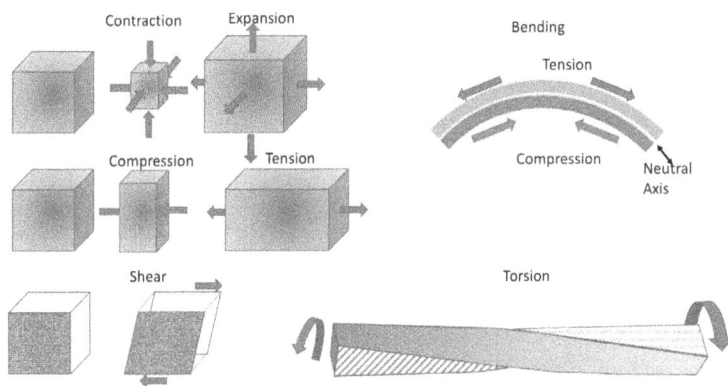

Figure. 12.3: Basic types of loads on structures

With ground-based structures such as bridges, the Margin of Safety is very large, in fact the yield stress may be 6 times the Maximum Permitted Stress. A Factor of Safety of 6. With aerospace structures we cannot afford the structural weight needed for such large factors. The Factor may be well under 2, although we will imagine that there will be events where the level of stress is far beyond the normal Design Stress. For instance, aircraft are typically designed for the high accelerations of maneuvers and landing. Thus the structure is built to take 3 times the weight, or designed to a 3G Load Factor. The landing gear may be designed for 9G or 12 G: we depend on it to take the extreme deceleration, and dampen it down to a far lower deceleration on the structure. Space vehicles are often limited to loads of less than twice the weight, perhaps 1.4 times the weight at liftoff. There are some large rockets which are in fact held up straight by the pressure inside their fuel tanks: remove all the fuel without supporting the structure, and it may collapse!

A spacecraft in its 3rd stage, accelerating to orbital speed, may be designed to accelerate at 3Gs. The decelerations of atmospheric re-entry are a different story: the vibrations there can be pretty violent. Let's leave that to someone else to worry about, for now.

Modern bridges are becoming more and more like airplanes, as structural designers *become more confident in the certainty of their predictions*. A corollary is that extreme events can collapse large bridges. A different side is that builders of bridges and skyscrapers are learning from Nature: their structures are now designed to deflect, float or sway with changing conditions rather than try to resist Nature. There are skyscrapers that are designed to sway with the wind, and even with earthquakes. We doubt if any can still survive the power of water as seen in tidal waves or tsunamis. But let's stay away from the temptation to bring in Disaster Movies.

12.2.1 Types of Loads

There are different kinds of loads, each causing a type of deflection. The basic types are tensile stresses and compressive stresses. The top left row in Figure 12.3 shows what happens to an object under uniform compression on all its faces (pressure) or uniform expansion as pressure decreases. The next row shows compression or tension occurring along only one direction. This causes a shortening or an increase in length. The third row shows a shear , caused by anti-symmetric tension or compression along surfaces. On the right, we see what happens with other combinations of stresses. When a rod or beam or plate is bent, the load is called a Bending Moment . The object deflects into a curved shape. The outside of the curve must have increased in length from the unstressed value. So there is tension along the outside. The inside shortens: there is compression there. The middle where the length has not changed, is the Neutral plane, sometimes called the Neutral Axis . How is strain measured? The classical way is using instruments called Strain Gauges. These are typically thin strands of wire set in epoxy and stuck to the surface where strain is to be sensed. A change in length of the wire causes a change in its electrical resistance, which is sensed by the electrical unbalance in the circuit to which it is connected. The circuit is typically based on something called the Wheatstone Bridge. Bending is sensed by attaching strain gauges to both the tension and compression sides of the object, and sensing the difference in resistances which is now doubled.

Figure. 12.4: Cable used to carry a slung load. Courtesy: US Army

Another combination of loads is shown below that: this is called Torsion . A rotation stress applied to the ends in opposite directions. The deflections resulting from such loads are called by corresponding names: "shear deflection", "bending deflection", and "torsional deflection". "Axial deflection" can refer to deflections caused either by compression or tension, so long as the load is acting in the axial direction. In the next figures, some examples of these deflections are shown. In practice, though the deflections are small, aircraft are large, so we can indeed see wings bending, the tip moving through several feet relative to the root. The wings are also under torsional loads. The wing tips may be seen to twist with their front edge pointing upwards or downwards (angle of attack increasing or decreasing) under aerodynamic load. This causes the wing to twist. The wings have to be designed to take this torsion as well.

Some values of strength and density are given in Table 12.1. You might conclude that Carbon Nanotubes sound very good, but we have not learned to make them very long (try a few centimeters) yet, so large structures that offer the full strength of these miracle materials, are a bit far off. On the other hand, the humble spider has learned to produce miracle materials: the silk of a spider's web is as strong as Titanium with a density that is less than 1/3 that of Ti. If we could get Prof. Spider to teach us Materials.

Material	Yield strength (MPa)	Ultimate Strength (Mpa)	Density kg/m_3
Mild Steel	248	841	7.58
Human skin	15	20	2.2
Titanium	940	1040	4.5
Aluminum alloy 2014-T6	414	483	2.8
Graphene	N/A	130000	1
Colossal Carbon Nanotube		3600	1.3
Spider silk		1000	1.3
Spruce wood		74/?	0.46
Tropical Balsa Wood		73/1	0.16
Carbon-Carbon Composite		700	1.98

Table 12.1: *Best Tensile Strengths of Materials. For Balsa wood, the values of tensile strength along/perpendicular to the grain are given.*

12.3   Some Basic Types of Structures

Some basic types of structures used in aerospace construction are cables, trusses, beams, shells and bulkheads. We will see the reason for using each type of structure, below. Trusses are very efficient structures, but do not make smooth surfaces. Monocoque structures (means single shell) have a thin shell with bulkheads, Semi-monocoque structures also have longerons.

You know what a cable is. It is a structure that can take only tension, but lots of it. Modern cables are often made of many individual strands of wire, wound around each other. The cable is very flexible (not much resistance to bending) and it cannot resist compressive loads along its axis, but it is quite stiff in tension, with the yield strength reaching that of the material of which it is made. Other cables today are made with a round tube, with a Kevlar web wound around it. The strength comes from the tensile strength of the Kevlar fibers, reinforced by the connections in the web. This is used in building the 6000-meter tethers used to hold large Lighter Than Air aerostats at 4000 meters above the ground in fairly strong winds.

Many bridges (I believe the Golden Gate Bridge in San Francisco, USA is one) have cables helping to hold up the bridge, tied to strong, tall towers. Several stunt movies have been made as these high-strung cables have captured the imagination of movie-goers, novelists and movie makers alike. Speaking of Disaster (Narrow Heroic Avoidance) Movies the Alistair MacLean novel "Goodbye California" and more than one Ian Fleming/ Albert Broccoli James Bond 007 movies surely build on the many possibilities. If you have walked on some Suspension (or Swing) Bridges across ravines that derivetheir support entirely from cables/ropes, you will prefer to look straight ahead and keep to the posted speed limit as you drive across such bridges. Somehow, being inside a car makes me feel safer than if I were walking on the bridge myself.

12.3.1 Truss

A truss is a structure composed of slender bars joined to other bars only through frictionless pins at their ends. This is one step above a cable because the elements can take both tension and compression, and the truss structure as a whole can take all sorts of loads. Trusses are amazing inventions, that support astonishingly large structures with seemingly featherweight elements. The bars are connected to external supports if any, only through end joints. Supports and loads must be applied at the joints. Since forces are applied at the ends. For equilibrium, the sum of forces at the ends, and the moments, must be equal and opposite. So the forces must be aligned with the bar. For any member in a truss, the member must be either in tension or compression, or have no force at all. In tension, members can take a large load. Pt = A.σ_y. In compression the same member can buckle at a much lower stress. Buckling stress $Pc = \frac{pi^2 EI}{L^2}$. Shorter members with the same cross-sectional area are harder to buckle. I is the moment of inertia of the member (or bar), the length is L and E is the material modulus of elasticity.

So in designing a structure, one minimizes the number of bars that are under compression, and makes sure that they do not buckle. In finite elements, this is done by iteration. We also use iteration to make sure that the loads are not too small (this would mean that the structure is too large and hence too heavy for the load).

A historic use of a variant/derivative/predecessor of trusses is in the wooden Trestle Bridges that enabled the American Railroad to well, 'railroad' its way across the canyons and ravines of the American West. OK, let me not dwell on the process of building any of those, they just amaze and terrify me. When completed they do look awesome in their beauty. Each trestle is defined as a rigid frame that can take some load, typically with a tripod construction.

The cool App "Truss Me!" (scientificmonkey.com/software.html) authored by Prof. Julian Rimoli of Georgia Tech works on both the Apple iOS and Android operating systems (a free download) teaches you about Structures through application. Structural analysis in practice is often done using the Finite Element method. Reproduced by permission, the discussion and table below set the stage to understand the philosophy of structural design. Linear and rotational motion have many parallels, and Table 12.2 below compares many of them. This should have been learned in basic physics.

$$a = \frac{F}{m}, \alpha = \frac{T}{I}$$

12.4 Moment of Inertia

The moment of inertia is to rotation what mass is to displacement. The more mass an object has, the more difficult it is to move. The more moment of inertia a structure has, the harder it is to rotate. The moment of inertia depends on both the mass of a structure and the distance of this mass from the axis of rotation; the farther the mass is from the axis of rotation, the larger

Linear	Rotational
force (F)	torque (T)
mass (m)	moment of inertia (I)
displacement (x)	angular displacement (θ)
velocity (v)	angular velocity (ω)
acceleration (a)	angular acceleration (α)

Table 12.2: Terms describing rotational quantities corresponding to linear ones

the moment of inertia. The I-beam, a basic structure often used in structures, derives its shape from this principle. With most of the mass located in the horizontal flanges, the I-beam resists rotation. A structure that uses these for support is less likely to undergo torsional deformation.

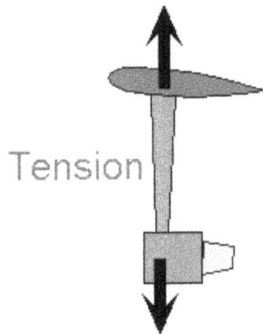

Figure. 12.5: Engine Strut: exaggerated representation

Next we will look at different types of structural members used to take the different types of loads. The engine mount takes the weight of the engine as shown in Figure 12.5, which causes tension, and the shear load (not shown in the figure) as the engine pulls the aircraft forward, or pulls it back during thrust reversal after landing. Of course the tensile and shear deflections of the engine mount are hardly noticeable, but you can in fact see the engines move slightly from side to side in torsional vibration when the wings vibrate during flight through disturbed air. Another way to take tensile load is using cables, see Figure 12.4. The initial airplanes had wings held together with cables, until people realized that nice airfoil shapes could reduce drag down far below what cables generated. The trusses that we discussed at the start of this chapter are elaborate structures that operate on the principle of tensile and some compression stresses.

Where compression loads are experienced, such as on the landing gear (see Figure **??**), columns with struts are used to prevent buckling.

Panels with webs are used to carry shear. Beams with spars are used to carry bending. Plates may carry compression and tension. Curved shells carry compression and tension and shear. The part of a beam that carries tension and compression is called a flange or a cap.

12.5 Aeroelasticity

Aeroelasticity refers to the study of structures under aerodynamic loading. Not only do aircraft encounter large aerodynamic loads when flying at high speeds, they have structures that are specifically designed to create yet more forces (like wings that create lift). Therefore it is very

Ribs for tension

Struts to prevent buckling in compression.

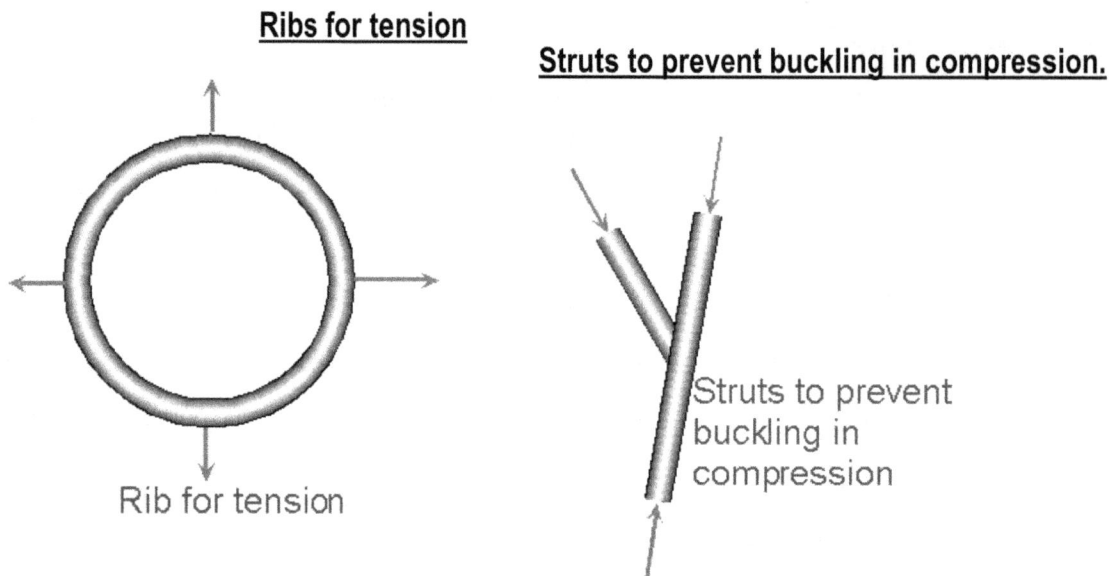

Rib for tension

Struts to prevent buckling in compression

Figure. 12.6: Ribs for tension and Y-struts for compression

important to see how structures and aerodynamic forces interact.

Figure. 12.8: View of an airliner wing in flight

An aircraft in flight is supported by the lift on its wings. The wings are structures that experience their own weight, the lift and drag distributed over their surfaces, and the pitching moment due to the lift. Let us imagine that these change in time, perhaps because the aircraft hit a region of disturbed air and the lift rapidly changed. The wing is made to be quite flexible, and it will bend and twist in vibrating response to the sudden excitation.

Looking out through the window, we can see the wing deflected significantly up in stead flight (see Figure 12.8). Going through turbulence, one can see the wing going up and down in bending. If one is an aerospace engineering professor one reassures oneself that every passing grade that one gave was well-deserved, and that the planes they design are very well built indeed. What happens to the air flow at the wing due to this 'plunging' motion? The angle of attack changes with time, and so does the lift. The wing can also twist due to the vibration, and again this can change the angle of attack. This is called a pitching motion. What happens, if the effect of the plunging motion is to increase the angle of attack in pitch, and that accentuates the plunging motion, and vice versa? The amplitude of both pitching and plunging could increase sharply, and so do the bending and torsional stresses. In less time than it takes to type this, the wings could break off! This is the feared phenomenon of aerodynamic flutter. Designers carefully avoid this, but sometimes it still hits us where we least expect it.

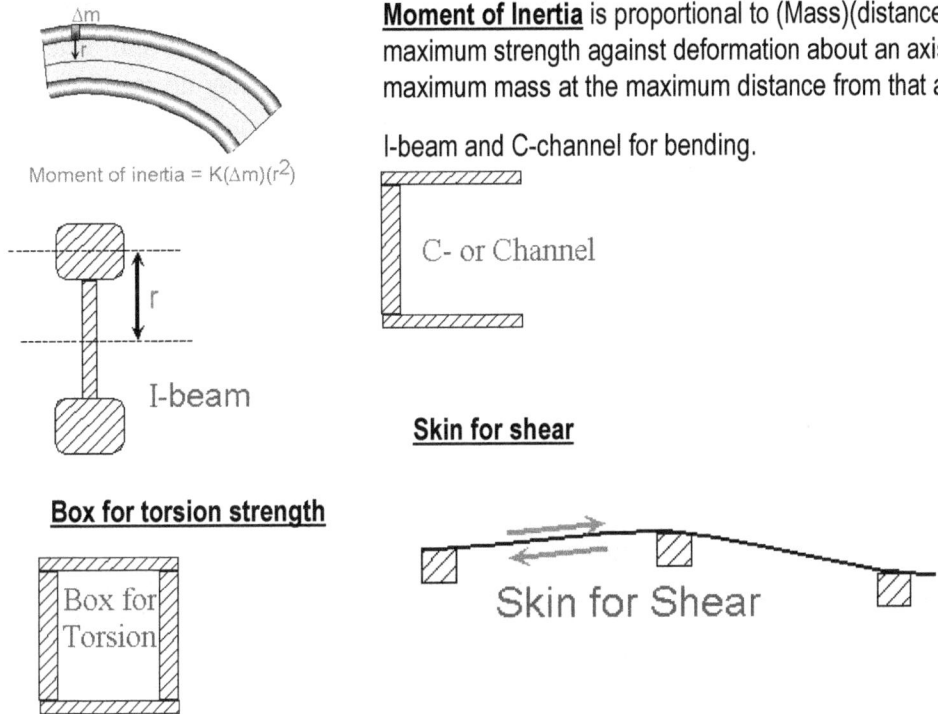

Moment of Inertia is proportional to (Mass)(distance)^2. So, for maximum strength against deformation about an axis, place the maximum mass at the maximum distance from that axis

Moment of inertia = K(Δm)(r^2)

I-beam and C-channel for bending.

C- or Channel

I-beam

Skin for shear

Box for torsion strength

Box for Torsion

Skin for Shear

Figure. 12.7: Structural elements to obtain high moment of inertia about desired axes. The top subfigure illustrates the idea of moment of inertia, applied to the bending of a beam. The I-beam is used to take symmetric bending. The C-channel is also used to take bending about two axes. The torsion box is used for strength in resisting torsion, for instance, placed in side the front part of an airfoil section. The skin with intermediate supports is used to take shear.

Figure. 12.9: The X-53 attempts to use small control surfaces to induce wing-warping and control the aircraft.

Flutter is a phenomenon in the field of dynamic aeroelasticity, and explains a rapid up and down vibration of the wings, and typically causes failure. It occurs when aerodynamic forces mix with elastic and inertial forces to create a dynamic instability. Although it can happen to bridges and chimneys, it is a huge design consideration when designing wings and other lift-creating surfaces. The vibration is usually caused by the aerodynamic force on the wings. If the vibration increases the aerodynamic force, the effect is self-exciting oscillation, and the wing undergoes rapid periodic motion before failing and breaking. Aeroelastic flutter is what caused the famous collapse of the Tacoma Narrows Bridge in 1940. Caused by 40 mph winds, the aerodynamic forces on the bridge induced an oscillation in the main span that eventually caused the bridge to collapse. Now aerodynamic forces are taken into consideration when designing bridges and other stationary structures. Even chimneys are designed with this in mind so that they do not flutter in high winds. The forces experienced by an aircraft in high-speed flight are much more severe, and for this reason aeroelastic effects are carefully considered when designing an aircraft.

Figure. 12.10: The Atlas family of rockets would reportedly collapse if the fuel tanks were mostly empty and the rocket were launched: the fuel provided the pressure to stiffen the structure. Credits: NASA

Until this point, aeroelastic phenomena have mostly been shown in a negative light. However, this does not have to be the case, and aeroelasticity's effects can be used to increase performance and control. This idea dates back to the Wright brothers, who, lacking the sophisticated control surfaces we take for granted today, tried to control their Wright Flyer by twisting and warping the wings. However, control surfaces such as ailerons quickly became the preferred method of aircraft control. The NASA Active Aeroelastic Wing program in recent times studied the feasibility of using aeroelastic effects to cause wing-warping in order to increase control and performance. The X-53 (the NASA "X" designation denotes 'Experimental') used to test the idea is a modified F-18 with multiple "aerodynamic tabs" meant to induce aeroelastic effects. The hope is that control of the aircraft can be achieved with smaller control surfaces that turn the entire wing into a control surface, rather than relying on a single large aileron. The larger ailerons add a lot of weight to the wing, and have minimum size requirements, due to the fact that they must be able to

control the aircraft even at low speeds. This causes a lot of strain at higher speeds, and also reduces the aspect ratio of the wing. By using the smaller control surfaces to induce twist in the wing, lighter, more efficient wings with higher aspect ratios can be used instead.

12.6   Weight

Aerospace vehicles must be as light as possible so they can perform as efficiently as possible. The most structural integrity is extracted from the least amount of material possible. This requires innovative structural designs not necessary in other fields. The implications of structural mass are perhaps most evident when considering rockets. The ideal rocket equation presented in the spaceflight chapter gives some insight into this. Achieving low earth orbit requires a velocity increment of well over 8,000 m/s. If we use some of the best fuel available, with a specific impulse I_{sp} of about 350 s, and an initial to final mass ratio of 10, the ideal rocket equation yields a velocity increment of only about 7,900 m/s. The structure and payload of the rocket only represent 10% of the initial rocket mass; the other 90% is the fuel! Although multistage rockets are used to circumvent this problem, this example makes very clear the need for highly efficient structures, capable of supporting many times their weight. The Atlas rocket would collapse under its own weight if the fuel tanks were empty. A structure so thin that it requires the pressure of fuel in the fuel tanks to keep it supported is precisely the kind of structural shenanigans that must be used in order to have successful vehicles.

The aircraft industry has primarily created airframes out of aluminum for many years due to its light weight and high strength. However, carbon fiber is a material that is both stronger and lighter. Carbon fiber is made up of thin strands of carbon. These strands are thinner than human hair, and can be woven together like yarn. When coated with some sort of resin, these strands can be molded into particular shapes. This is similar to papier mache, which involves coating newspaper strips with glue so that they mold to some shape.

Being both lighter and stronger than aluminum, carbon fiber has attracted the attention of commercial aircraft designers. The Boeing 787 is 50% carbon fiber by weight and 80% carbon fiber by volume. This represents a major advance in commercial aviation. The aircraft is estimated to be 20% more fuel efficient than a similarly sized aircraft that is built with the more traditional aluminum.

Although there are many great advantages to using carbon fiber, there are also many hurdles impeding its use. While fractures and imperfections in aluminum are easy to test for (in many cases being visible to the naked eye), it is much more difficult to test and perform maintenance on carbon fiber. Carbon fiber is also more expensive than aluminum, but is becoming cheaper with time.

In this chapter we have seen a very succinct description of a massive field of aerospace engineering. Consider that the vast majority of work in building airplanes and rockets is, of course structures. With new materials being developed, there are immensely exciting possibilities. For instance, except for the composite structures briefly alluded to at the start of this chapter, most structures are still made of metal, with is an *isotropic* substance: the strength is the same along all directions inside the metal.

Figure. 12.11: The Boeing 787 is the first commercial airliner made mostly of carbon fiber. Credits: Boeing Co.

But with composites, and increasingly with new 'designer materials', engineers can decide how to distribute the available mass in the best way, to obtain immense strength along the directions that they want the structure to resist, at incredibly low cost in mass. Imagine the resulting airplanes. Today it takes a minimum of 25 percent of the gross takeoff weight of a large airliner, just to build the structure. In future perhaps this could be cut to 10 percent, leaving so much more for payload. To develop such structures, engineers must learn to delve into fundamental properties of materials down to the subatomic level, learn about nano-material properties, and perhaps also spend more time watching the real experts of Nature build their structures: birds building nests, and spiders weaving webs.

Chapter 13

High Speed Flight

13.1   Introduction

Low-speed flows can be considered incompressible. This means that density is assumed constant. However, this assumption cannot be made for high-speed flows. This makes calculations for high-speed flows more difficult.

13.2   Mach Number and Mach Angle

13.2.1   Mach Number and the Speed of Sound

Small disturbances in air propagate at the speed of sound. This means that the speed of sound is the speed through at which information propagates through the fluid. The speed of sound, a, can be calculated with the formula below:

$$a = \sqrt{\gamma R T}$$

where γ is 1.4 for air, T is an absolute temperature (Kelvin or Rankine), and R is the gas constant of air.

The Mach number M is defined as:

$$M \equiv \frac{u}{a}$$

13.2.2   The Mach Cone Angle

The Mach Cone defines the boundaries of where the weakest disturbances created by an object moving faster than sound, reach, around and behind the object. As shown in Figure 13.1 the

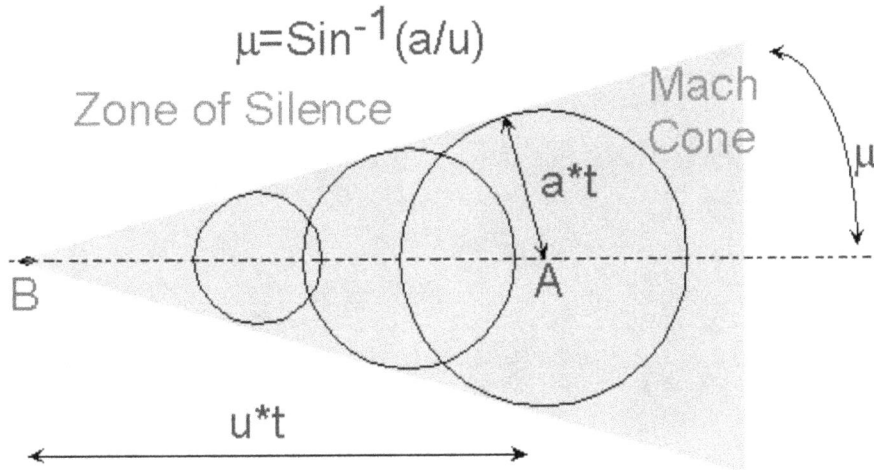

Figure. 13.1: MachCone

Mach cone is the result of the disturbance spreading outwards in an expanding sphere radiating out from the source, as the source moves faster than the speed of sound. Note that the picture shown in the figure is an instantaneous picture. At the next instant, the object will have moved, and the cone will have grown in size in the window that we are viewing.

$$\sin \mu = \frac{a}{u} = \frac{1}{M}$$

$$\mu = \sin^{-1} \frac{1}{M}$$

13.3   Shocks

In aeronautics, the term "shocks" refers to shock waves. These waves develop in front of bodies traveling faster than the speed of sound. Because the speed of sound is the speed at which information propagates through a fluid, air in front of an object traveling faster than the speed of sound is unaware of the impending disturbance. The shock develops to slow down the flow before it strikes the object. Across a shock wave, the Mach number of the fluid decreases, and huge increases in pressure, density, temperature, and entropy occur.

If a body is blunt, the shock wave forms in front of the body, as shown in the left side of Figure 13.2. If the body is sharp, an oblique shock wave forms.

Oblique shocks turn the flow before it hits the object. The oblique shock angle β is greater than the Mach angle μ.

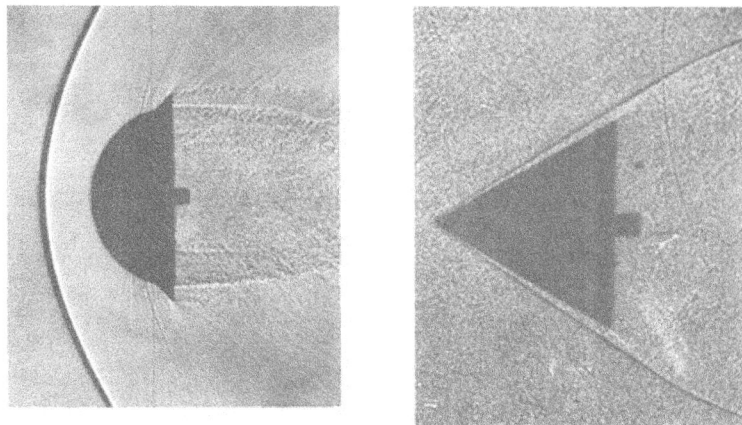

Figure. 13.2: (Left)A bow shock wave. (Right)An oblique shock wave. Credits: NASA

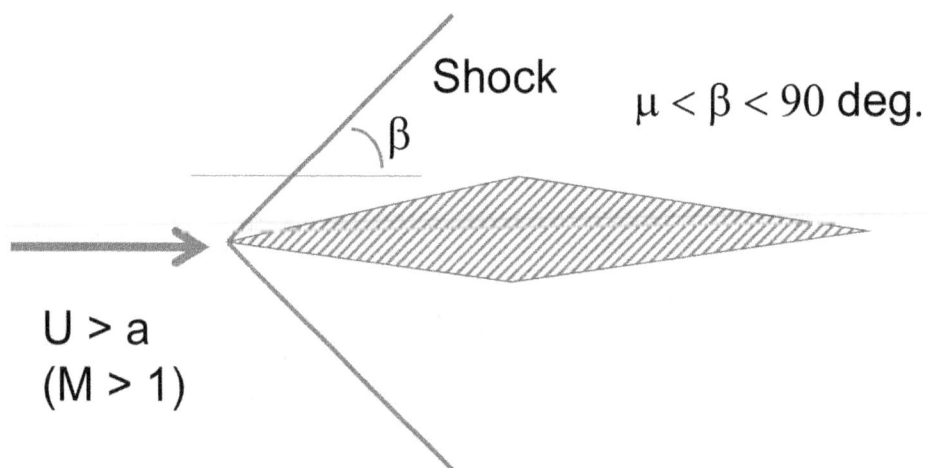

Figure. 13.3: Oblique shock angle β.

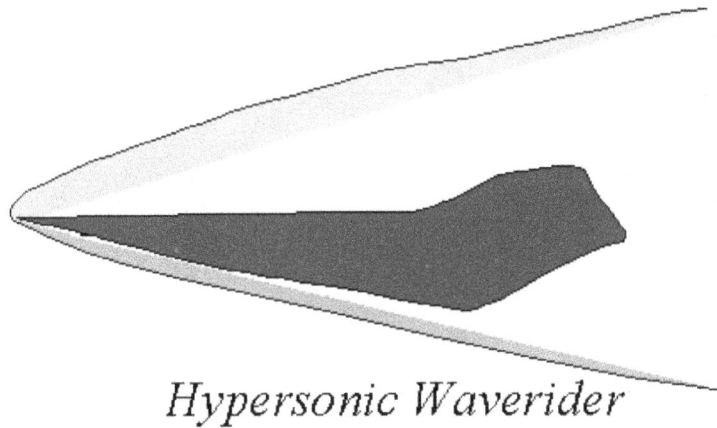

Hypersonic Waverider

Figure. 13.4: NASA hypersonic flight test with an airbreathing supersonic combustion ramjet (SCRAMJET) engine. Credits: NASA

13.3.1 Hypersonic Regime

The hypersonic regime exists above Mach 4 or 5. Across a shock there is a large pressure increase, as well as a large temperature increase. The Mach angle is extremely small. It is possible to generate lift with a large pressure rise below the lower surface of a hypersonic aircraft.

13.3.2 The Power of Shocks

If an asteroid is so big that it still remains as a solid object when it nears the ground (i.e., comes down to a few thousand feet), the shock waves can hit the ground with terrible strength. This can flatten trees, forests, and buildings (happened in Siberia just around the time the Wright Brothers were getting off the ground). When the object finally reaches the ground, the shock in front blasts a huge crater in the earth's surface. The crater is huge, even though the object that actually reaches the ground may be quite small. The Meteor Crater in Arizona is one mile in diameter, and hundreds of feet deep. It is widely believed that an immense object hit the Gulf of Mexico during the age of the dinosaurs, and the havoc was such that it caused the extinction of the dinosaurs, worldwide.

13.4 Flight at High Subsonic and Transonic Speeds

Above a Mach number of about 0.3, air can no longer be considered incompressible. This makes calculations more difficult, as density can no longer be considered a constant. One of the smartest moves I have seen in all of engineering is the Subsonic Transformation technique, where the behavior of a flow, and the pressures on surfaces in flows, can be calculated at any subsonic Mach number, using the results obtained at any one Mach number. This isolates the effects of

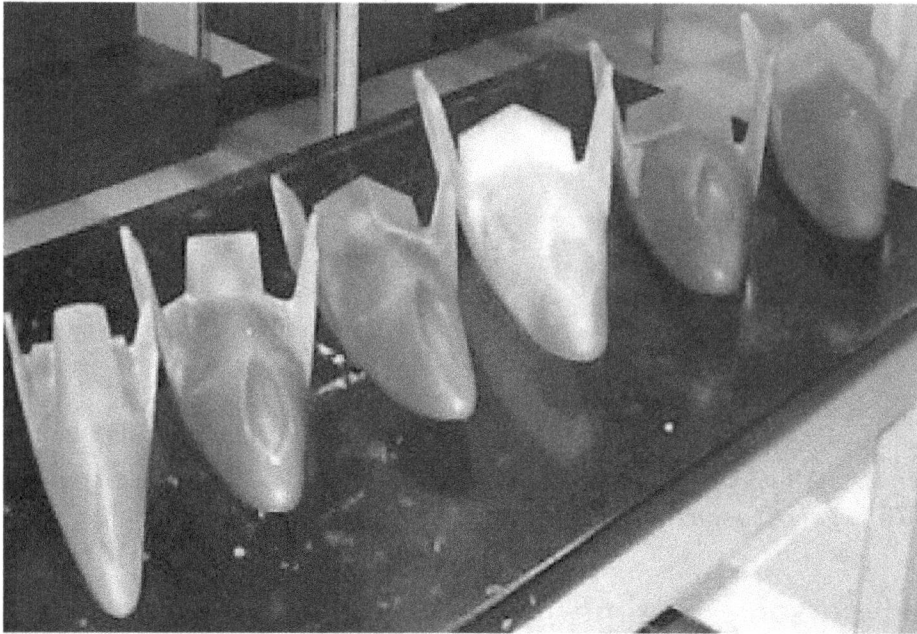

Figure. 13.5: Several different models of NASA X-38 Crew Return Vehicle configurations prepared for tests at the John J. Harper wind tunnel. Courtesy Experimental Aerodynamics Group, Georgia Institute of Technology.

compressibility - as distinct from speed, density or dynamic pressure. Both analysis and testing are extremely difficult and expensive, in the subsonic regime if you try to do it directly. The transformation technique saves us all that trouble. Usually the condition at which we have, or can obtain, results is the incompressible regime where Mach number is well below 0.3. Testing, or simple analysis, in this regime can be used to go all the way up in Mach number to the edge of the transonic regime, where supersonic flow first appears on a configuration. In this course we will see only one result from that transformation technique: the Prandl-Glauert formula.

13.4.1 Subsonic Flight: Prandtl-Glauert Compressibility Correction

Information about an airfoil's lift and drag coefficients in incompressible flow is not valid in the compressible regime. However, the incompressible lift and drag coefficients can be converted into lift and drag coefficients in the compressible regime using the Prandtl-Glauert compressibility correction. The lift and drag coefficients of an airfoil at a given angle of attack increase with flight Mach number, according to the Prandtl-Glauert Compressibility Correction:

$$c_l = \frac{c_{lo}}{\sqrt{1 - M_\infty^2}}$$

$$c_d = \frac{c_{do}}{\sqrt{1 - M_\infty^2}}$$

where c_{lo} and c_{do} are the lift and drag coefficients before the correction. To get the same lift coefficient at a higher Mach number, one requires a smaller angle of attack. This is good, because

The Critical Mach Number

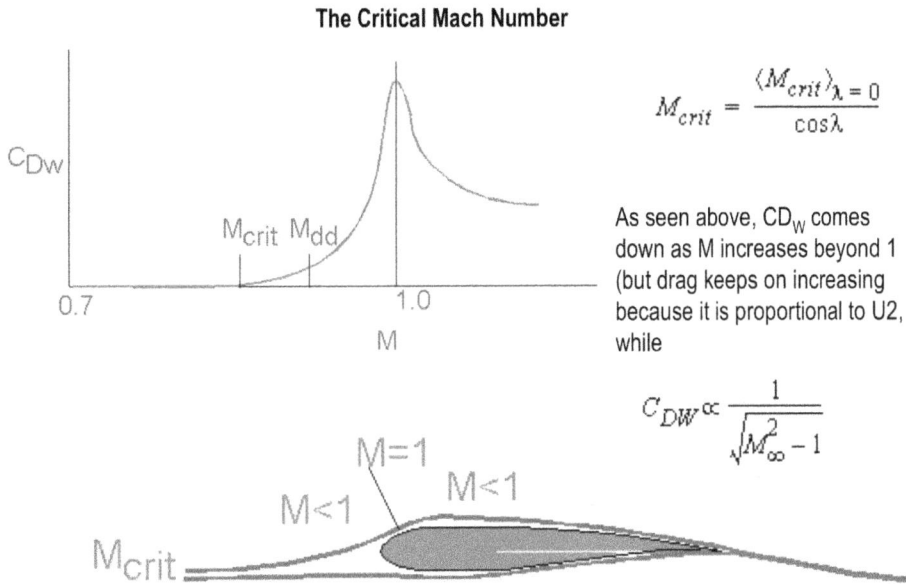

$$M_{crit} = \frac{\langle M_{crit} \rangle_{\lambda = 0}}{\cos\lambda}$$

As seen above, CD_W comes down as M increases beyond 1 (but drag keeps on increasing because it is proportional to U2, while

$$C_{DW} \propto \frac{1}{\sqrt{M_\infty^2 - 1}}$$

Figure. 13.6: Drag increases greatly near Mach 1.

the drag coefficient also increases like this. This expression is valid for Mach numbers which are lower than the critical Mach number.

13.4.2 Critical Mach Number

As a wing approaches Mach 1, the flow at some point along the wing will reach Mach 1 even if the freestream Mach number is lower. Once any flow along the airfoil reaches Mach 1, huge increases in drag occur, as the graph below shows. For this reason, commercial aircraft fly just below their critical Mach number, so a huge drag increase can be avoided.

It is possible to "trick" the flow with swept wings and achieve higher critical Mach numbers. The motivation to do so should be apparent; an increase in the critical Mach number is effectively an increase in the cruise speed. The most notable method is the use of wing sweep. If a wing is swept back an angle Λ, the wing only experiences the component of the freestream normal to it. This component is $u_\infty \cos\Lambda$. Because this normal component is smaller than the freestream velocity, the aircraft can travel at a higher freestream velocity before the flow over the wing at any point reaches Mach 1. Thus, the critical Mach number is higher with wing sweep. The positive effects of wing sweep can be seen in commercial aircraft. The table below demonstrates the relationship between the sweep angles of various airliners and their cruise Mach numbers.

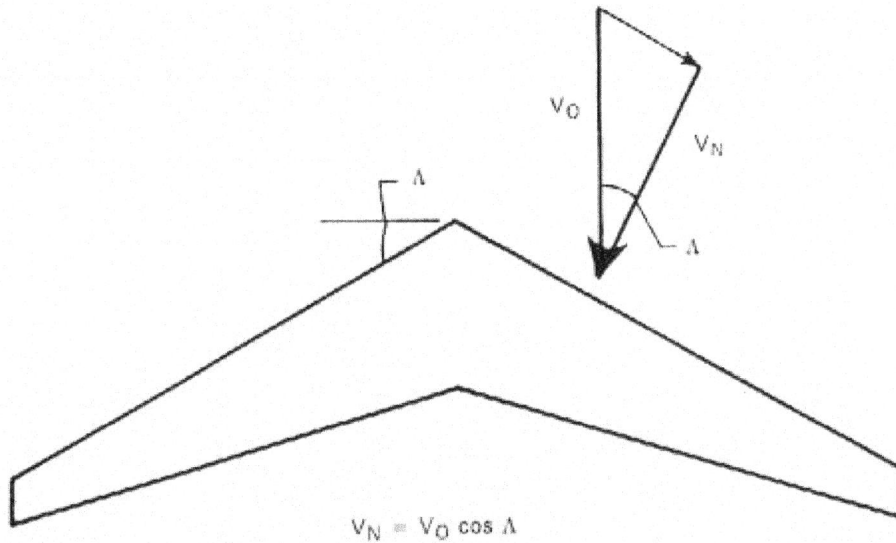

Figure. 13.7: Wing sweep increases the critical Mach number by decreasing the normal component of velocity experienced by the wing.

Aircraft	Sweep Angle (°)	Typical Cruise Mach	Max Cruise Mach
B737-800	25.0	0.785	0.82
B757-200	25.0	0.80	0.82
B767-300	31.5	0.80	0.86
B777-200	31.64	0.84	0.89
B747-400	37.5	0.85	0.92

13.4.3 Suggested Method to Calculate Transonic Drag Peak

The part up to Mach 1 can be described as a Lorentz function.

$$C_D = \frac{1}{\pi\gamma[1 + (\frac{M - M_{cr}}{\gamma})^2]}$$

Where at the peak

$$C_{Dmax} = \frac{1}{\pi\gamma}$$

The initial attempts to fly supersonic with human-carrying aircraft and airbreathing engines, encountered difficulties in the transonic regime because the engines had barely enough power to get through the transonic drag peak. Some fighter aircraft of the late 1950s/ early 1960s could exceed Mach 1, but only in shallow dives. There was a Convair design (F-102) that encountered surprisingly high drag due to shock interactions. The fuselage was redesigned using the famous 'Coke Bottle' shape, constricting the fuselage diameter to compensate for the cross-sectional area of the wings. This allowed the aircraft to go supersonic. The transonic area rule became part of standard aircraft design. Today the thrust-to-weight ratio of jet engines has gone far higher than in those days, so punching through the transonic drag rise is not a problem. Modern fighter planes do not appear to conform to the transonic area rule. This is because they are designed to go much faster than Mach 1, and the "area ruling" may become evident when one considers the area intercepted by Mach cones at the design Mach number with the shape contours. The rise of 'stealth' as a design criterion also pushed aerodynamic design down in priority. Designing a supersonic aircraft for minimum drag, with quick iterations between shapes, continues to be a strong challenge because of the difficulty in predicting shock-shock interactions and shock-corner flow interactions accurately. The point of the above paragraph is that computing the transonic drag peak is probably not very important in the initial stage of designing a modern supersonic aircraft.

13.5 Example Problems

1. Now that you are in the final stages of completing a conceptual design of a flight vehicle, please write briefly (say half a page), comparing what you have learned, with the reasons you gave in Assignment 1 which you submitted near the beginning of the semester. How has your idea of aerospace engineering changed / been confirmed? What areas do you find interesting now, versus then? If the answer is ÒI only like Space, so this does not applyÓ, well, no, you canÕt get away with that. Comment on the aero part. Even spacecraft have to take off and land (if they ever return) through the atmosphere. I will compare your answer to your Assgt 1 submission; you donÕt have to have memorized anything, but surely you remember the sentiments you expressed then.

2. A new supersonic airliner that causes very little disturbance, is coming towards a bird. The aircraft is traveling at Mach 2, at an altitude where the temperature is 218K. As the nose

glides by, 30 meters directly below the bird, the bird looks down, wondering why she canŎt hear anything from this shiny monster. How many milliseconds after the nose passes, will the bird feel the first disturbance? Explain with a diagram and calculations.

3. An airfoil is found to have a lift coefficient of 0.2 at a certain angle of attack, when tested at Mach 0.1. Estimate its lift coefficient at Mach 0.7 at the same angle of attack.

4. You are designing the inlet of a jet engine. At the design Mach number of 2.0, the first disturbance from the lip of the inlet must touch the roof of the inlet (see diagram). Find the ratio of distances h/L.

13.6   Example 2: Supersonic Airliner Design

In this section we will go through a (rudimentary!) conceptual design of a hydrogen-fuelled Supersonic Transport and illustrate what is needed to 'close the design'. First let us summarize the parameters of the Concorde, the only supersonic airliner to have conducted international operations for decades.

The source is *cs.mcgill.ca/ rwest/wikispeedia/wpcd/wp/c/Concorde.htm*

With the parameters given below, we will use an EXCEL spreadsheet to try to show that the *design closes* for our hydrogen-fuelled Supersonic Transport.

13.6.1   Specified Parameters

- Range: 7000 miles (convert to kilometers)

- Passengers 250+

- Baggage: at least 10 metric tons (10,000 kg mass).

- Crew: 10

- At least 28% of the takeoff weight is needed to build the aircraft (Structure Fraction).

- The propulsion system (engines and associated items but not the fuel) will have a thrust-to-weight ratio at takeoff, of 10.

- Supersonic cruise lift-to-drag ratio is 7.

- Cruise altitude is 18288 m (60,000 feet) and Mach number is 1.7.

OK, let us get started.

- At this altitude, static Temperature is 216.5K

Parameter	Value
TakeOff Weight TOW:	188,000 kg (413,600 lbf)
Planform Area S:	358.25 square meters
W/S takeoff :	5143 N/m^2 (112.74 psf)
W/S cruise:	4436 N/m^2 (92.5 psf)
Fuel weight fraction:	0.55
Crew:	9
Passenger capacity:	100
Length:	202 ft 4 in (61.66 m)
Wingspan:	84 ft 0 in (25.6 m)
Height:	40 ft 0 in (12.2 m)
Fuselage internal length:	129 ft 0 in (39.32 m)
Fuselage max external width:	9 ft 5 in (2.88 m)
Fuselage max internal width:	8 ft 7 in (2.63 m)
Fuselage max external height:	10 ft 10 in (3.32 m)
Fuselage max internal height:	6 ft 5 in (1.96 m)
Wing area:	3,856 ft? (358.25 m?)
Lift/drag ratio: Low speed-	3.94
Approach	4.35
250 knots, 10,000 ft	9.27
Mach 0.94	11.47
Mach 2.04	7.14
Empty weight:	173,500 lb (78,700 kg)
Useful load:	245,000 lb (111,130 kg)
Powerplant: 4? Rolls-Royce/ SNECMA Olympus 593 Mk 610 afterburning turbojets	
Dry thrust (each engine):	32,000 lbf dry (140 kN)
Thrust with afterburner (each engine):	38,050 lbf (169 kN)
Performance	
Maximum speed:	Mach 2.04 (1,350 mph, 2,170 km/h)
Range:	3,900 nm (4,500 mi, 7,250 km)
Service ceiling:	60,000 ft (18,300 m)
Rate of climb:	1,525 m (5,000 ft)/min (25 m/s)
Wing loading: (Takeoff)	5143 N/m^2 (112.74 psf)
Wing loading: (cruise)	4436 N/m^2 (92.5 psf)
Thrust/weight:	0.373
Fuel consumption:	46.85 lb/mi (13.2 kg/km)
Maximum nose tip temperature:	260 deg. F (127 deg. C)

Table 13.1: *Comparison of a liquid hydrogen vs. hydrocarbon fuelled SST*

- Pressure is 7171.64 $\frac{N}{m^2}$

- Viscosity is 1.43226E-5 in SI units.

- Find density using the perfect gas law as $\rho = \frac{P}{RT}$

- Find the speed of sound as $a = \sqrt{1.4RT}$ where 1.4 is the value of γ, the ratio of specific heats of air.

- Assume that the engines will have the same thrust at cruise altitude of 60,000 feet and cruise Mach number of 1.7, as they do at the sea-level static case. (This is feasible with engine design for supersonic cruise conditions).

- So your wing loading can be increased a bit.

- Assume that one cannot operate at more than 80 percent of the maximum thrust during cruise (you donÕt want to burn out the engines early).

- The thrust-specific fuel consumption with hydrogen fuel is 0.18 per hour at cruise. Why is this so low? Because hydrogen releases nearly 4 times as much heat as petroleum fuels when burned in air.

- You can use the Breguet Range Equation as given in the Performance chapters with L/D constant.

Please be sure to calculate, in sequence:

1. Speed of sound

2. Flight speed

3. Dynamic pressure at cruise altitude and Mach number

4. Payload mass

5. Payload weight

6. Guess Takeoff Weight assuming liquid hydrogen fuel

7. Assign Propulsion weight fraction

8. Estimate installed Sea Level Static Thrust.

9. Estimate fuel weight.

10. Assume that the ADDITIONAL fuel consumed during climb to cruising altitude (note: you already calculate fuel burn for the whole range assuming cruise speed) is 5% of the total fuel weight. So assume that only 95% of total fuel weight is available for the cruise phase. The descent and landing phase is OK since you will consume less fuel, but we wonÕt take any credit for that.

11. Estimate average aircraft weight during cruise.

12. Guess cruise wing loading.

13. Find wing area

14. Limit wing span to 66m. Find aspect ratio

15. Find cruise lift coefficient

16. Estimate cruise drag

17. See if 80 percent of the maximum sea-level static thrust is enough to overcome the drag in cruise.

18. If not, reduce wing loading and try again.

19. Given the range, find the fuel weight needed.

20. Check if there is enough weight available to build the structure.

21. Iterate: change parameters so that there is only just enough weight to build the structure.

22. Obtain the lowest-weight aircraft that does the job.

23. Give a name to your aircraft company, and give the aircraft a Make and Model Number.

24. Report the takeoff gross weight of the aircraft, along with your other parameters: range, cruise Mach number, number of passengers, cargo mass in metric tons, weight fractions of payload, fuel, engines and structure, number of engines and maximum sea-level static thrust per engine.

Parameter	Value	
SST Drag Calculation		
Parameter	Liquid Hydrogen	Hydrocarbon
altitude, m	14000	14000
Mach	1.7	1.7
Temperature, K	216.5	216.5
Density, $\frac{kg}{m^3}$	0.226753	0.226753
Viscosity, Pa.s	1.43226E-05	1.43226E-05
Speed, m/s	502	502
Reynolds Number per meter	7.95E+06	7951871
Dynamic Pressure q, Pascals	28602	28602
$\sqrt{M^2-1}$	1.3747727	1.3747727
Re	3.90E+08	3.98E+08
$Cd_Mach1.7$ (Boeing Correlation)	0.001176	0.001176
Payload	301840	301840
Payload fraction	0.3	0.15
TOW, N	1006133	2012267
CruiseWt	855213	1710427
W/S	2400	2400
S	356	713
Span	40	60
AR	3	3
Av. Chord	8.9	11.9
CL	0.083909654	0.083909654
Cdlift	0.002419885	0.002419885
DragDueToLift	24664	49327
L/D_{lift}	35	35
Fuselage Cylinder dia, m	5.75	5.8
Length, effective cylinder	49	50
FusArea	885	911
Aircraft length for Sears-Haack	81	81
Volume, m^3	1272	1321
Wing Volume, 3% thick	254	
Total Volume	1368	1575
VolumeWave Drag	43829	47244
L/D_{volume}	20	36
Total area	1598	2336
Friction Drag	53745	78588
Total Drag	122237	175160
L/D	7.00	9.76494
Fuel Fraction	0.25	0.55
Fuel Mass	25666.66667	112933.3333
Fuel Density	70	800
Fuel Volume, m^3	366.6666667	141.1666667
Fuel Fraction of total vol	0.268104258	0.089629942

Table 13.2: Supersonic Transport Aircraft Drag Estimation: LH2 vs. HC

Chapter 14

Space Flight

14.1   Rockets Revisited

Figure. 14.1: The glare of the Sun behind the Lunar Excursion Module of Apollo 14, at the surface of the Moon. Courtesy NASA.

Velocity Increment ΔV: Consider a rocket with effective exhaust velocity c_e. The effective exhaust velocity is a way of expressing the thrust in a simple manner by adding up the momentum thrust and the pressure thrust and dividing this sum by the mass flow rate. As propellant is blasted from the exit nozzle, the mass of the vehicle decreases. This is substantial in the case of the rocket as compared to air-breathing engines, because all the propellant comes from inside the vehicle. From Newton's Second Law,

$$\text{Thrust} = M\frac{dV}{dt} = -c_e\frac{dM}{dt}$$

$$dV = -c_e\frac{dM}{M}$$

$$\Delta V = -c_e\int_{M_1}^{M_2}\frac{dM}{M}$$

$$\Delta V = c_e\ln\frac{M_1}{M_2}$$

where M_1 is the initial mass, which includes the propellant, and M_2 is the mass after the propellant has been used up to achieve the velocity increment ΔV .

Specific Impulse and Mass Ratio Define the specific impulse of the propellant as:

$$I_{sp} \equiv \frac{c_e}{g}$$

163

where g is the standard value of acceleration due to gravity at sea-level (9.8 m/s^2). The unit of specific impulse is seconds. Using this definition,

$$\Delta V = gI_{sp} \ln \frac{M_1}{M_2}$$

Mass ratio of a rocket is

$$\frac{M_1}{M_2} = e^{\frac{\Delta V}{gI_{sp}}}$$

Note: Some space agency websites express specific impulse without the g. Thus their "I_{sp}" is simply c_e.

14.2   The Laws of Newton and Kepler

Newton's Law of Gravitation To find the velocity increment required for various missions, we must calculate trajectories and orbits. This is done using Newton's Law of Gravitation:

$$F_r = -\frac{GMm}{r^2}$$

Here the left hand side is the "radial force" of attraction due to gravitation, between two bodies; the larger is of mass M, and the smaller is of mass m.

The universal gravitational constant is:

$$G = 6.670 \text{ x } 10^{-11} \text{ Nm}^2/\text{kg}^2$$

Keplers Laws These can be applied to a satellite of mass m, orbiting a much larger object of mass M (m $\ll$ M).

1. The satellite travels in an elliptical path around its center of attraction, which is located at one focus of the ellipse. The orbit must lie in a plane containing the center of attraction.

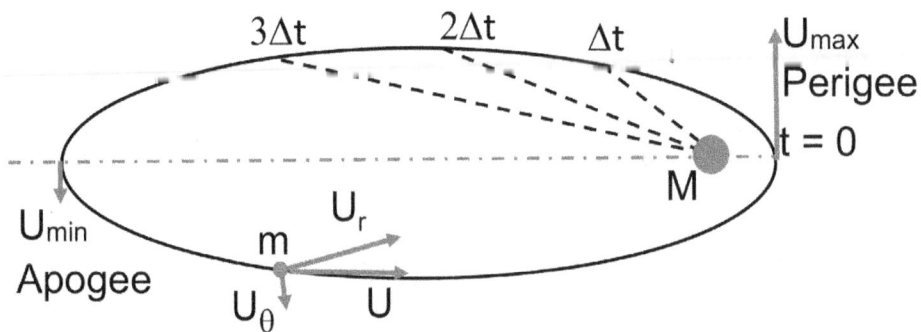

Figure. 14.2: Illustration of Kepler's Laws.

2. The radius vector from the center of attraction sweeps equal areas of the orbit per unit time. As the satellite moves away, its speed decreases. As it nears the center of attraction, its speed increases.

3. The ratio of the squares of the orbital periods of any two satellites about the same body equals the ratio of the cubes of the semi-major axes of the respective orbits.

$$\frac{t_1^2}{t_2^2} = \frac{a_1^3}{a_2^3}$$

Calculating Velocities Along an Elliptical Orbit The speed at any point in an elliptical orbit is:

$$v = \sqrt{g_0 r_0^2 (\frac{2}{r} - \frac{1}{a})}$$

where g_0 is the acceleration due to gravity at radius r_0.
The minimum velocity is obtained at the apogee , or highest point of the orbit:

$$v_{apogee} = \sqrt{g_0 r_0^2 \frac{(1 - \varepsilon)}{a(1 + \varepsilon)}} = v_{min}$$

where eccentricity is given by

$$\varepsilon = 1 - \frac{r_{min}}{a}$$

The max velocity is obtained at the perigree , or lowest point of the orbit:

$$v_{perigree} = \sqrt{g_0 r_0^2 \frac{(1 + \varepsilon)}{a(1 - \varepsilon)}} = v_{max}$$

The time for one orbit is:

$$\tau = \frac{2\pi\sqrt{a^3}}{r_0\sqrt{g_0}}$$

Circular Orbit and Escape Velocity at Radius r

$$v_{circ} = \sqrt{g_0 \frac{r_0^2}{r}}$$

$$v_{escape} = v_{circ}\sqrt{2}$$

$$v_{escape} = \sqrt{2g_0 \frac{r_0^2}{r}}$$

Example: Speed for Circular Orbit and Speed for Escape
Earth's volumetric radius $\approx$ 6370.998685023 km ($\approx$ 3958.755 mi; $\approx$3440.064 nmi). However, note that in calculating the standard value of g_0, a slightly different radius is used:

$$g = G\frac{m1}{r^2} = (6.5742 \times 10^{-11})\frac{5.9736 \times 10^{24}}{(6.37101 \times 10^6)^2} = 9.822\frac{m}{s^2}$$

Consider a circular orbit at r = 7000 km.

$$v_{circ} = \sqrt{g_0 \frac{r_0^2}{r}}$$

Escape speed at radius r.

$$v_{escape} = v_{circ}\sqrt{2}$$

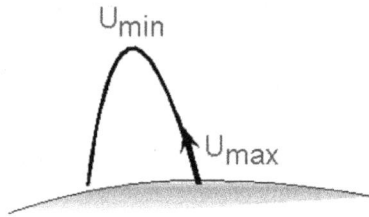

Figure. 14.3: Launching a projectile. Perigee and Apogee concepts applied to a ballistic trajectory.

A Projectile in a Ballistic Trajectory You see orbital mechanics at play when you throw a ball, or when a long-range missile is launched. Or when you jump up. In each of these, the launched object is going into an orbit around Earth, except that the trajectory is a bit too narrow, the eccentricity is too high and the energy too low, for it to escape hitting the Earth as it returns from Apogee. This is seen in the figure below. If you throw a ball straight up, the speed at apogee is zero and it falls straight back down. Where the initial speed is very high and it is launched at an angle, the curvature of Earth becomes an issue in calculating the range. In general, the Dynamics books tell you that the best angle to launch, for maximum range, is 45 degrees. You see from long-exposure photos of night-time Space launches, that this is basically what they do. At the very start they lift straight up with a very small inclination angle to clear the launch area, and then they tilt over gradually until the angle is not far from 45 degrees. In the scheme of things, considering the huge range, the time for which the rocket engines operate (the Burn Time) is very short, more of less like your arm coming arcing over and straightening out as you throw. If the speed is high enough, and the angle is large enough, the projectile will indeed go into an orbit around Earth.

14.3   Calculating a Space Mission: The Space Shuttle

It is quite easy to calculate the progress of a space mission. A bit of careful geometry is needed, true, but otherwise it is all straightforward application of Newton's Laws.

1. We can use the Rocket Equation to quickly estimate the rough Mass Ratio for a given mission. Given the mass remaining at the end, one can then directly estimate the mass at the outset. Let us then start with this mass, and an assumed vehicle. Unlike the aircraft design, we will go directly to the performance estimation, with firm guesses of the technologies used for the fuel, the structure, and the engines.

2. At the start, the thrust must become greater than the weight, for the rocket to lift off. How much greater? Depends on the mission. For a small missile, particularly Surface to Air missiles, getting up to altitude as fast as possible is all-important, so the thrust-to-weight ratio will be quite high, and the acceleration will rapidly exceed 1 G. For a giant vehicle such as the Space Shuttle, the structural weight fraction has to be kept low. The excess of thrust over weight at the start is only from 5 to 15 percent, so that acceleration is at most 0.15G. But then as the propellant is used up, and the engines keep generating the same thrust, the speed rises, and so does the acceleration.

3. As soon as the thrust develops, a small inclination is applied to the direction of the thrust, so that the rocket moves slowly away from the support structure. In other words the vehicle has a small but increasing sideward velocity component. This is largely left to the designers to specify. The nose is also tilted a bit so that it is not much away from the trajectory (otherwise angle of attack would become large). Again, there are various ranges of values to use here.

4. We calculate the balance between thrust, drag and gravitational force at time t=0, hold the values constant and calculate how far the vehicle travels, how much the angles change etc within a very short interval. Say at 0.1 second. We repeat the calculation, with the vehicle now having a finite value of velocity components as well as acceleration. We keep repeating this at small time intervals (the beauty of a spreadsheet is that you can make the computer do the work by simply pulling down on the cells down the columns).

5. In the case of the Space Shuttle, a strange problem is encountered. The speed rises so fast that at some altitude, the density is still high enough, that dynamic pressure exceeds some critical value. This is called Max Q in NASA parlance. Beyond this level of dynamic pressure, if there are fluctuations due to gusts etc,

the structure might be at risk of damage. The thrust of the main liquid-propelled engines is throttled back to keep the acceleration from getting too high, allowing the vehicle to climb beyond this critical zone.

6. At some point the booster solid rockets are burned out. In the spreadsheet , we sharply reduce the vehicle weight by the weight of the empty boosters, and keep going.

7. We keep track of the velocity, the height and the direction of the velocity. The idea is to have all tangential velocity by the time we reach orbit.

8. The curvature of the Earth must be accounted, to accurately gauge all the vectors and the altitude above the ground. This requires elementary geometry, knowing the radius of the Earth. Accounting for the spin rate of the Earth allows us also to track where we have reached, above what point on Earth. In general, space missions are launched towards the East, because the Earth is moving that way too, so we get the advantage of the tangential speed of Earth. What matters for orbit is the energy of the orbit, which comes from the tangential velocity and radius with respect to the center of the Earth. This additional help from Earth's rotation is quite substantial, being close to the speed of sound! You can calculate it: the Earth has a rotation period of 24 hours and the radius can be obtained from the Atmosphere chapter.

All said and done, our students found that they reached orbit at just about the time that the real Space Shuttles did, jettisoning the Main tank just before orbit so that the tanks would come back down and crash into the ocean.

14.4   Other Matters of Interest

One could easily write another 200 pages on the prospects, glory, dangers, and all other aspects of Space science and engineering. What aerospace engineer can look up into the blue sky and not wonder about the infinite distances along that line of sight? What we have discussed above is an extremely brief introduction to the basics. Consider that Earth is just a tiny, tiny part of the Solar system, which is a sort of unremarkable stellar system, one about a billion stellar systems, moving along somewhere in the middle reaches of the Milky Way Galaxy, one of billions of Galaxies in the part of the Universe that we can ever hope to sense with the best of telescopes. Most of today's large telescopes are a far cry from Galileo's or Kepler's tubes of polished metal and glass lenses. Today we capture information sensed over the entire range of frequencies of electromagnetic radiation: from very very long wave waves that require sensors spaced over the entire planet to capture. For example, the Long Interferometer gizmo LIGO (I doubt if that is the official expansion of the acronym) uses 3 points spread out across the planet to capture gravitational waves. Large arrays of antennae in the Chilean desert and other remote places (reduced interference from human activity and weather) capture somewhat shorter wavelengths of radio waves (typically frequencies of hundreds of KiloHertz out to many MegaHertz). There are telescopes that operate in the Microwave regiem as well (1 GigaHertz to over 10 GigaHertz). Astronomical observatories, some located atop high mountains such as the volcanos of Hawaii at 2800 meters above sea level, capture much of the spectrum from the Ultraviolet, through the Infrared, out to Millimeter Waves (100 to 300 GigaHertz).

What we see from time to time, published from NASA and other organizations, are often composites of the information extracted from these (plus, I suspect, a liberal amount of artistic liberty, from the best Space Artists of our times).

See? I wrote a long paragraph already and I am not even scratching the surface of what I wanted to describe. All of the above is part of Aerospace Engineering, although much of this is done as part of large Physics projects, involving huge teams of scientists, engineers of all disciplines, and people to make those cool artistic renderings, and the best of electronics innovation. Much of today's advances in image and signal processing technologies, came to the mass market starting from military projects, through programs such as SETI (Search for Extraterrestrial Intelligence) and Space Telescopes. Keeping the big Space Telescopes (Hubble, Spitzer, Chandra) aligned precisely in tiny spots of the Universe, even as they hurtle around Earth (how else do they stay where we can communicate with them?)

Space is, well... ultimately spacious. The numbers are all mind-boggling. How can one even begin to wrap one's brain around the idea of Space? Place some limits? I found an essay by Isaac Asimov [131] to be a tremendous breakthrough in my own understanding and comfort with Space concepts. He showed that there are ranges of frequency, and relative speeds that have finite bounds. Some important ideas of science bind together. One is the relationship between wavelength (a measure of length), frequency (a measure of rate, inversely proportional to time), and speed of propagation. I seem to remember an argument that placed a bound of 75 kilometers per second on the speed of any object that we were likely to see hurtling past Earth (the big ones that actually hit, you won't ever see). Similarly there was some limit on frequency that ran out of room beyond the regime of X-rays because the wavelength became too short. There was probably a limit on the longest wavelength, but I do not remember that.

Scientists look to Space to test hypotheses on the laws of Physics. They seek to refine all-inclusive Grand Models of Physics. One often reads arguments from Physicists who seem to be much more comfortable declaring how things are in some galaxy far, far away, than about things in front of our eyes or in the next town. I wonder if they really know those things for sure. This is surely a continuing reason to explore Space: we still know very little of the realities beyond Earth. But there is a far more important (to me) reason to explore: this is to build economic growth and development in a responsible and thoughtful manner.

A book by Beatty *et al* [132] presents recent discoveries about the places in our Solar System. Lewis [4] goes into detail about the prospects of finding valuable resources, particularly in the Asteroid Belt, and among the Near-Earth Objects which are asteroidal masses in orbits much closer to Earth and in some cases following Earth in the same orbit. Seager [90] writes about the atmospheres that may be expected to be found on exoplanets - planets orbiting stars other than the Sun. Much closer to home, there is increasing interest in the characteristics of planetary atmospheres in our Solar System, most importantly about Mars [91, 92] and Venus [133–135]

14.5   Example Problems

Space colony design We are considering designing a space habitat to be placed at the L-5 point of the Earth-Moon system. The habitat will be designed for 100,000 people. New research has shown that virtually all healthy humans can easily withstand an environment rotating at 1.2 RPM, and that they can live without any ill effects at 0.7Gs. Please compute the diameter and length of a circular cylinder shell for this application, minimizing the radius. If this habitat is to have a radiation shield 2m thick, made of lunar regolith (lunar top soil), what is the mass of the radiation shield, given that the specific gravity of regolith is 2.3?

The Personal Space Vehicle Trips to the Space Mall Ghwynnett are to be done using the GM2020 PSV (Personal Space Vehicle), which can be used from other locations in Eath Orbit. The highest deltaV capability (you can refuel at the Space Mall if you need the fuel for the return trip) is 2km/sec. What is the specific impulse needed if the Mass Ratio is to be kept below 1.8? If necessary, assume that only transfers between circular orbits are considered.

Lunar Base The velocity increment needed for a given space mission (reach a good lunar orbit from the surface of the Moon) is 2.2 km per second. If the vehicle at fuel burnout is to have a mass of 2000 kg, and the specific impulse of the propulsion system is 400 seconds, what is the initial mass of the vehicle at launch from the MoonÕs surface? Given that gravitational acceleration at the Moon's surface is 1/6 that on Earth's surface, what is the fuel mass flow rate needed to produce enough thrust, so that the vehicle has a net upward acceleration of 15 $\frac{m}{second^2}$ in the very first second of flight after takeoff.

14.6 Project Example: Rocket Hot Fire Testing

What is Hot Firing of a rocket? The idea is that many tests of the principles of rocket engines are done with cold flow, that is pressurized gas blowing out through a convergent-divergent nozzle. The tests below are real rocket engines, although very small rocket engines. The following example is taken from an assignment in the AE1601 course, provided by various instructors.

Background Information Teams of 3 students (typically) design a model rocket using an open source software application called OpenRocket. Each team is given a target apogee altitude, and performs an iterative design process using OpenRocket to develop a rocket design to reach the target altitude as closely as possible. Idealized thrust profiles for selected model rocket engines are available from the engine manufacturers. In addition, the hot-fire testing generates experimental thrust profiles, enabling comparison of simulation with experiments.

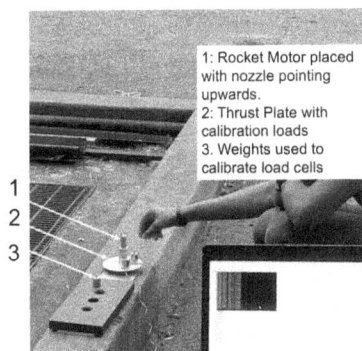

Figure. 14.4: Model Rocket engine thrust stand being calibrated before testing

Figure 14.4 shows a small model solid rocket motor being readied for testing. Of course, note that you cannot test a model solid rocket motor without using it up, so there is no question of re-using it. The data that you obtain will give a sample set of data, and you expect that rockets motors that are of the same type, from the same company, and the same batch of manufacturing, will exhibit a thrust-time history that is very similar to the one that you obtained. More on this below. The Space Shuttle's Solid Rocket Boosters are recovered, but even those must be refurbished carefully for any re-use. The small model motors seen here, have paper wrapping, and the nozzle is merely a shape formed into the propellant grain during manufacture. The figure shows an engineer placing known weights on plate that is placed above the load cell (which cannot be seen). The laptop computer screen at right bottom shows the data being collected from the load cell, through an analog-digital converter. As the loading weight is increased, the measured voltage also increases. Plotting this curve shows the calibration of the load cell.

The calibration weights are then removed and the motor is ignited by passing a high current through the wires seen leading up to the motor. The current is high enough to burn the igniter wire inside the combustion chamber that is formed inside the solid rocket grain. This decomposes the propellant near the wire, and the gases that are released, react with air, releasing far more heat. Ignition!

The hot gas rapidly fills up the combustion chamber. The rate at which gas is released, exceeds the rate at which they can escape through the small throat of the nozzle, which is thus choked . In other words the Mach number of the gases reached 1 at the throat, even using the speed of sound of the gases at that high temperature. This causes the pressure inside the combustion chamber to build up sharply, which drives up the flow going through the throat. The speed at which the gases blow out, thus rises rapidly. So does the thrust.

And then the nozzle bursts due to the high pressure inside, and is blown off. Suddenly the nozzle throat is the diameter of the rocket motor's casing. From then on the motor becomes an end-burning motor, which means the burn rate is constant. Because the nozzle throat is gone, the acceleration to supersonic speeds is also gone; the thrust is now much lower. The motor continues at this thrust, until the propellant is all exhausted.

OK, you have now obtained actual data on a Solid Booster Static Test. Start with that.

1. Plot out the calibrated thrust in Newtons vs. time.

2. From that, find the Total Impulse.

3. From the difference between the initial weight and the weight after the propellant is gone, estimate the

mass of propellant. It should be in the neighborhood of 12 grams.

4. From this, find the Specific Impulse. Convert to units of seconds using the factor $9.8 \frac{m}{sec^2}$ as needed. You should get something in the vicinity of 60 to 90 seconds.

5. Also find the Average Thrust. Ignore the negative thrust at the start when there is some unsteadiness: start averaging only when the thrust really goes up to stay up.

6. End the averaging when the thrust drops sharply just before extinction: do not keep averaging the long near-zero tail.

7. Find the Average Steady-State Thrust: average only the part after the initial big spike in thrust has subsided, and before the thrust drops at the end.

8. Now look for the peak thrust reached. That is when the chamber pressure was highest. Shortly thereafter, the nozzle formed from propellant gets burned/blasted out, leaving you with just a hollow tube which is a pretty dismal combustion chamber (very low pressure) and even worse nozzle. So! Pick the thrust at the peak. This should be perhaps 4 times the Average Steady State Thrust Or maybe a bit more than 4 times. This is the thrust you would have had, if your nozzle had stayed on, and the combustion chamber could take the pressure.

9. Multiply the Specific Impulse by the ratio between Peak Thrust and Average Steady State Thrust. This is the Specific Impulse that you might have had if the nozzle and combustion chamber had stayed intact. In a real rocket booster we would build these things in. Hope you get something in the range of 200 to 300 seconds. If not, well, check with someone: we should not be getting 100, or 800! Anything over 300 is probably wrong. With 100, the Shuttle people should have got their money back ?

Example: Rocket Design The goal of this phase is to design a two-stage model rocket that, when "flown" via a trajectory simulation, reaches a specified target apogee altitude. The OpenRocket program allows the design and trajectory simulation of model rockets through a user-friendly interface. First, download the OpenRocket program from the website: http://openrocket.sourceforge.net/

Open the program on your computer and try working it. The website also contains documentation and a Wiki page. To familiarize yourself with OpenRocket, you may first want to develop a single-stage rocket design that includes all of the necessary elements, then move to a two-stage design. The User Guide accessible through the Wiki page has a Basic Rocket Design set of instructions that walks you through an example: $http : //wiki.openrocket.info/Basic_{R}ocket_{D}esign.$

Your rocket design should have a nosecone, body tube, fins, launch lug (this is the tube that the launch rod passes through), payload (described below), shock cord and parachute recovery system, an inner tube (to house the engine), an engine block, and centering rings (to center the inner tube). The payload for the rocket flight is an instrument package. In our case it is usually an altimeter plus data recording system. The PerfectFlite Pnut altimeter is described at the site $http : //www.perfectflite.com/pnut.html$. The altimeter dimensions there are 63.5 mm x 15 mm x 11.4 mm, and the mass is 7.3 g. The altimeter measures static pressure, and records altitude as a function of time. The altimeter must be mounted within the rocket tube, and it should be returned to Earth via the recovery system. (Note: wadding is also considered part of the payload mass in OpenRocket. The mass of the wadding should be accounted in addition to the altimeter mass shown above).

The next step is engine selection. OpenRocket has a wide range of engines that are pre-defined, and you can select the desired engine(s) from a menu. One company that sells model rocket engines is Estes Rockets Inc. $https : //www.estesrockets.com/$. No endorsement implied but that is the only one whose website I have visited. Estes rocket engine specifications. For the two-stage rocket design, you may choose from the following Estes booster engines and upper-stage engines.

- Booster Engine Options: B6-0, C6-0

- Upper Stage Options: A6-2, A8-3, B4-4, C6-5

In a two-stage design, the most reliable approach is direct staging, where the upper stage is placed directly above the booster engine in a stack. When the booster stage burns through its black powder propellant, hot gases and burning particles from the booster will ignite the upper stage. In practice, the two stages are usually joined by a single wrap of cellophane tape (you do not have to model this). The tape serves to hold the engines together during upper stage ignition, but the tape is melted by the hot gases and it allows the stages to separate following upper stage ignition. Once the rocket is designed and the engines are selected, you can then use OpenRocket to simulate the launch trajectory. For this project, we will assume zero wind velocity.

Through iterative combinations of booster and upper stage options, you can choose an engine configuration that allows you to approximate the apogee altitude target. Also ensure that your vehicle is stable (center of pressure below center of gravity), and that the recovery system allows a survivable surface impact velocity. OpenRocket allows you to assess these design parameters.

Example: Rocket Engine Testing Once the Design Phase is complete, you will inform the instructor what booster and upper stage rocket engines you have selected. During hot-fire test each rocket engine type will be tested and thrust profiles will be measured using a load cell. The raw test data are sent to the entire class.

The next step is to simulate trajectory performance with measured thrust profiles. From the test data repository, you will retrieve the thrust profiles for your booster engine and upper stage. From this data, you need to create a custom thrust curve text file that can be read into OpenRocket. The thrust curve file template is given at the website: http://www.thrustcurve.org/raspformat.shtml. Basically, the body of the thrust curve file contains the thrust as a function of time from ignition.

Within OpenRocket, you can add your custom thrust curve to the OpenRocket database of engines through Edit/Preferences/Options. Where it says "User defined thrust curves:" click the ADD button. Browse to your thrust curve test file and click ADD. Once the custom thrust curve has been selected, you must quit and restart the OpenRocket application. When it is restarted, the custom thrust curve will appear in the database of engines. Redesign your rocket to use the measured thrust profiles from the hot-fire testing. Simulate the trajectory performance with the measured thrust profiles, and compare the trajectory against that for the ideal thrust profiles.

As specified in our classes, final reports should capture the design process, specify the final design and engine selection, document the hot-fire test results, and compare the trajectories of the rocket using the measured ideal engine thrust profiles. In the design specification, you should provide enough detail to allow a reviewer to replicate your results in OpenRocket. Project reports should utilize the AIAA Technical Journal format. Reports should be printed on 8.5 x 11 inch paper, stapled in the corner. No fonts smaller than 12 point.

14.7   Example 3: Space Launch

This example is to see why the Space Shuttle used to do what it used to do. Now for the real rocket. Look up the parameters of the Space Shuttle. How many solid rocket boosters (SRB) did it have? What was the starting thrust of each? Weight? How much of the weight was propellant?

Based on actual data for the SRB, the average thrust of each was around 2.6million pounds force (sorry, you will have to convert that to Newtons). They burned for about 110 seconds (they burned a couple of seconds longer, but that was used to get them away from the Shuttle. It is not too smart to hold on to an empty rocket casing after the propellant is gone, so they would have erred on the conservative side and ejected the booster before it was totally empty and extinguished.

The Main Engines (liquid hydrogen/liquid oxygen or LH2/LOX for us Rocket Scientists) could be throttled to vary thrust.

So! Let us start the calculation. Get the EXCEL sheet ready, put down all the constants etc that you need.

1. Find total takeoff weight, and the total takeoff mass. You need both.

2. Find initial total thrust. You will be underestimating the boosters by using the average value I gave above, but do not worry. If you seriously want to use more accurate time-varying thrust values you can find them on the web, but they are given for sea-level; they will indeed change with altitude. Not that easy to calculate.

3. Find the initial acceleration from Newton's second law (you know the total mass, you know acceleration due to gravity which acts downward, you know the thrust acting upward, so you know the net force upward. Divide that by mass and you know the acceleration.

4. Assume that it accelerates at that rate for 0.1 second. Find the upward speed reached, and the distance travelled in 0.1 second.

5. At this point, the upward speed is the only component of velocity.

6. Find the new density for the new altitude reached.

7. The drag coefficient is based on the cross-section area of the Space Shuttle (which is mostly the cross-section area of the main tank + about 10% of that as a guess). Assume that this drag coefficient is around 0.12. Constant all the way to orbit. Find the drag force.

8. The drag is directed along the flow velocity vector.

9. Find the resultant force acting on the vehicle due to thrust, weight and drag. This is the acceleration vector.

10. Find the amount of propellant expended from both the solid and the liquid engines.

11. Find remaining total mass.

12. Repeat steps 3 to 6 in 0.1-second intervals until you reach about 2 seconds.

13. At this point, tilt the thrust like 1 degree so that you have a small thrust acting sideways.

14. From this point on, calculate the sideways acceleration and speed reached, and the distance, as well as the vertical acceleration and speed and distance.

15. Calculate the new position (vertical and horizontal).

16. Keep up this process at least until the trajectory angle becomes horizontal. Note that even if the trajectory looks horizontal from your reference point, it is moving away from the surface of the Earth since the Earth is round, so that the actual altitude is increasing. You can worry about this later.

17. Keep track of the dynamic pressure "q", and the acceleration level in "Gs". Read about the "Max q" problem in the Shuttle launch trajectory. As you reach this region, you have to keep reducing the main engine thrust so that you donÕt exceed the "Max q".

18. As the boosters are released (close to their burnout point) the total mass suddenly goes down. You may have to reduce the main engine thrust so that the acceleration level stays below 3 Gs.

19. Now try plotting the trajectory from your reference point, and see how that looks superposed on the shape of the Earth's surface. From this, calculate the actual altitude reached, and the tangential speed reached.

20. The objective is that at about the orbit of the Space Shuttle, the tangential speed should reach a level where the centrifugal force balances the centripetal force which is gravitational acceleration.

21. In reality, G also changes a bit. You can calculate that easily, knowing the Gravitational Constant and the radius from the center of the Earth. So the orbital speed needed is a bit less than that calculated assuming constant G.

22. When you reach there, you can shut down the main engines: you have reached orbit and are quite "weightless".

23. See how well your numbers compared with published numbers for the Space Shuttle. You should not be very far off.

There are many other "special features" such as angle of attack, thrust vectoring, thrust variation with altitude, pressure thrust, aerodynamic lift, variation of drag with Mach number, shock-boundary layer interactions, effect of compressible boundary layers, hypersonic Real Gas Effects, etc etc. I think you should come pretty close to the real numbers with the above. Feel free to adjust the drag coefficient a bit: it may get high in the transonic phase (between Mach 0.85 and 1.1).

In this first edition of this book, we will not attempt to give the full listing of numbers and plots involved in the above. It can get daunting to see, but if you follow the directions above, you will see that you are also generating spreadsheets with lots of numbers, and intricate plots. This is far beyond what most people believed that they could do as first-year engineering students, which is the whole point of this book: your brain is a wonderful resource. Given some good motivation and discipline which people have already tried to encourage inside you, you can achieve a great many things. The book does show you how to estimate the limits, while showing that anything within those limits can be achieved.

Please do develop that large-capacity supersonic airliner with the spacious seats and leg-room and ticket prices that are within my ability to buy!! As for the Space Launches, I will be happy to read in the newspapers when your design flies or you fly on them. Read too much on Rocket Instability and more to the point, on why the Vomit Komet got its name....

All the best in your education and far beyond!

Bibliography

[1] P. Yarrow, P. Stookey, and M. Travers, "Leaving on a jet plane," Peter, Paul and Mary, Album 1700, 1967.

[2] S. Miller, "Jet airliner," Steve Miller Band, Book of Dreams Album, 1977.

[3] M. Gardner, *Great essays in science.* Pocket Books, 1957.

[4] J. S. Lewis and R. A. Lewis, *Space resources: Breaking the bonds of Earth.* Columbia University Press, 1987.

[5] M. Crichton, *Timeline.* Ballantine Books, 2013.

[6] J. Anderson, *Introduction to Flight*, ser. McGraw-Hill Series in Aeronautical and Aerospace Engineering. McGraw-Hill Higher Education, 2005. [Online]. Available: http://books.google.com/books?id=Hd_AR0CAmsoC

[7] R. Shevell, *Fundamentals of flight.* Prentice-Hall, 1983. [Online]. Available: http://books.google.com/books?id=N3hTAAAAMAAJ

[8] H. Tennekes and H. Tennekes, *The Simple Science of Flight: From Insects to Jumbo Jets.* Mit Press, 2009. [Online]. Available: http://books.google.com/books?id=lt4PQPDhX5YC

[9] A. M. Lindbergh and C. A. Lindbergh, *Listen! the wind.* Harcourt, Brace, 1938.

[10] M. Simons, *Model aircraft aerodynamics.* Nexus Special Interests, 1999. [Online]. Available: http://books.google.com/books?id=R2xvegHiRecC

[11] D. Alexander and S. Vogel, *Nature's Flyers: Birds, Insects, And The Biomechanics Of Flight*, ser. Nature's Flyers. Johns Hopkins University Press, 2004. [Online]. Available: http://books.google.com/books?id=zj395mz_GYkC

[12] D. Alexander, *Why Don't Jumbo Jets Flap Their Wings?: Flying Animals, Flying Machines, and How They Are Different.* Rutgers University Press, 2009. [Online]. Available: http://books.google.com/books?id=XCk2OwAACAAJ

[13] S. Brandt, *Introduction to aeronautics: a design perspective.* Aiaa, 2004.

[14] T. Damon, *Introduction to Space: The Science of Spaceflight*, ser. Orbit A Foundation Series. Krieger Publishing Company, 2011. [Online]. Available: http://books.google.com/books?id=K3IAPwAACAAJ

[15] K. Kemp, *Destination Space: Making Science Fiction a Reality.* Virgin, 2007. [Online]. Available: http://books.google.com/books?id=rlBGAAAAYAAJ

[16] U. S. C. A. Patrol, *Introduction to Flight.* National Headquarters, Civil Air Patrol, 1972. [Online]. Available: http://books.google.com/books?id=NRYbRAAACAAJ

[17] P. Hill and C. Peterson, *Mechanics and Thermodynamics of Propulsion*, 2nd ed. McGraw-Hill New York, 1991, p. 760.

[18] W. J. Hesse and N. V. Mumford, *Jet Propulsion for Aerospace Applications.* Pitman Pub. Corp., 1964.

[19] G. C. Oates, *Aerothermodynamics of aircraft engine components*.   Aiaa, 1985.

[20] ——, *Aerothermodynamics of gas turbine and rocket propulsion*.   Aiaa, 1997.

[21] S. Murthy, *High-speed flight propulsion systems*.   Aiaa, 1991, vol. 137.

[22] J. D. Mattingly, *Elements of gas turbine propulsion*.   McGraw-Hill Science, Engineering & Mathematics, 1996, vol. 1.

[23] N. Cumpsty, "Jet propulsion (a simple guide to the aerodynamic and thermodynamic performance of jet engines)," 1997.

[24] A. F. El-Sayed, *Aircraft propulsion and gas turbine engines*.   CRC Press, 2008.

[25] S. Farokhi, *Aircraft propulsion*.   John Wiley & Sons, 2014.

[26] W. H. Heiser and D. T. Pratt, *Hypersonic airbreathing propulsion*.   Aiaa, 1994.

[27] H. A. Spang III and H. Brown, "Control of jet engines," *Control Engineering Practice*, vol. 7, no. 9, pp. 1043–1059, 1999.

[28] C. D. Brown, *Spacecraft propulsion*.   Aiaa, 1996.

[29] W. R. Sears, *Theoretical aerodynamics*.   WR Sears., 1951, vol. 1.

[30] I. Abbott and A. Von Doenhoff, *Theory of Wing Sections, Including a Summary of Airfoil Data*, ser. Dover Books on Physics and Chemistry.   Dover Publications, 1959. [Online]. Available: http://books.google.com/books?id=DPZYUGNyuboC

[31] L. M. Milne-Thomson, *Theoretical aerodynamics*.   Courier Corporation, 1966.

[32] H. Ashley and M. Landahl, *Aerodynamics of wings and bodies*.   Courier Corporation, 1965.

[33] W. Vincenti and C. Kruger, *Introduction to Physical Gas Dynamics*.   Krieger Publishing Company, June 1975.

[34] P. A. Hanle and T. Von Kármán, *Bringing Aerodynamics to America*.   The MIT Press, 1982.

[35] J. J. Bertin and M. L. Smith, *Aerodynamics for engineers*.   Prentice-Hall, 1998.

[36] J. J. Bertin, *Hypersonic aerothermodynamics*.   AIAA, 1994.

[37] J. Anderson, *Fundamentals of aerodynamics*.   McGraw-Hill New York, 2001, vol. 2.

[38] E. L. Houghton and P. W. Carpenter, *Aerodynamics for engineering students*.   Butterworth-Heinemann, 2003.

[39] M. Drela, *Flight Vehicle Aerodynamics*.   MIT Press, 2014.

[40] D. J. Peery, *Aircraft structures*.   Courier Corporation, 2011.

[41] A. K. Noor, *Structures technology: historical perspective and evolution*.   AIAA, 1998.

[42] T. L. Lomax, *Structural loads analysis for commercial transport aircraft: theory and practice*.   Aiaa, 1996.

[43] T. H. G. Megson, *Aircraft structures for engineering students*.   Elsevier, 2012.

[44] O. A. Bauchau and J. I. Craig, *Structural analysis: with applications to aerospace structures*.   Springer Science & Business Media, 2009, vol. 163.

[45] I. Chopra and J. Sirohi, *Smart structures theory*.   Cambridge University Press, 2013, vol. 35.

[46] A. Rao, *Dynamics of particles and rigid bodies: a systematic approach*.   Cambridge University Press, 2005.

[47] C.-T. Sun, *Mechanics of aircraft structures*.   Wiley-Interscience, 1998.

[48] J. J. Wijker, *Spacecraft structures*.   Springer Science & Business Media, 2008.

[49] B. Lawson, *How designers think: The design process demystified*.   Routledge, 2006.

[50] D. Ullman, *The mechanical design process.* McGraw-Hill Higher Education, 2015.

[51] D. P. Siewiorek and R. S. Swarz, *The theory and practice of reliable system design.* Digital press, 1982.

[52] J. Roskam, *Airplane Design: Preliminary calculation of aerodynamic, thrust and power characteristics*, ser. Airplane Design. Roskam Aviation and Engineering Corporation, 1985. [Online]. Available: http://books.google.com/books?id=bo9TAAAAMAAJ

[53] W. J. Larson and J. R. Wertz, "Space mission analysis and design," Microcosm, Inc., Torrance, CA (US), Tech. Rep., 1992.

[54] J. Rifkin, *The hydrogen economy: The creation of the worldwide energy web and the redistribution of power on earth.* Penguin, 2003.

[55] D. Shukla, N. Hiremath, and N. Komerath, "A cycloidal rotor and airship system for on-demand hypercommuting," SAE Technical Paper, Tech. Rep., 2016.

[56] J. E. Bridges, C. A. Brown, B. S. Henderson, and J. A. Seidel, "Nasa's pursuit of low-noise propulsion for low-boom commercial supersonic vehicles," in *2018 AIAA Aerospace Sciences Meeting*, 2018, p. 0265.

[57] W. F. Milliken, "Equations of motion: Adventure, risk and innovation," *Warrendale, PA: Society of Automotive Engineers, 2006. 680*, 2006.

[58] I. Newton, "Philosophiae naturalis principia mathematica, vol. 1-4," *Colonia: A. Philibert; 548 p.; in 8.; DCC. 4.221 I through IV*, 1760.

[59] J. L. Meriam and L. G. Kraige, *Engineering mechanics: dynamics.* John Wiley & Sons, 2012, vol. 2.

[60] R. W. Soutas-Little and D. J. Inman, *Engineering mechanics: dynamics.* Prentice Hall Upper Saddle River, NJ, 1999.

[61] A. Pytel and J. Kiusalaas, *Engineering mechanics: dynamics.* Nelson Education, 2016.

[62] I. Jong and B. Rogers, *Engineering mechanics: dynamics.* Oxford University Press, 1990, vol. 2.

[63] I. H. Shames, *Engineering Mechanics: Dynamics.* Prentice-Hall, 1966, vol. 2.

[64] W. F. Riley and L. D. Sturges, *Engineering mechanics.* Wiley, 1996.

[65] R. C. Hibbeler, *Engineering mechanics.* Pearson education, 2001.

[66] S. P. Timoshenko and D. H. Young, *Engineering mechanics.* McGraw-Hill, 1956.

[67] B. Y. White, "Designing computer games to help physics students understand newton's laws of motion," *Cognition and instruction*, vol. 1, no. 1, pp. 69–108, 1984.

[68] J. McCarthy, "Situations, actions, and causal laws," DTIC Document, Tech. Rep., 1963.

[69] B. Ellis, "The origin and nature of newton's laws of motion," *Beyond the edge of certainty. Englewood Cliffs, NJ*, pp. 29–68, 1965.

[70] R. J. Blackwell, "Descartes' laws of motion," *Isis*, vol. 57, no. 2, pp. 220–234, 1966.

[71] S. S. Antman and J. E. Osborn, "The principle of virtual work and integral laws of motion," *Archive for Rational Mechanics and Analysis*, vol. 69, no. 3, pp. 231–262, 1979.

[72] D. W. Sciama, "On the origin of inertia," *Monthly Notices of the Royal Astronomical Society*, vol. 113, no. 1, pp. 34–42, 1953.

[73] A. Einstein, *Relativity: The special and the general theory.* Princeton University Press, 2015.

[74] K. S. Thorne and J. B. Hartle, "Laws of motion and precession for black holes and other bodies," *Physical Review D*, vol. 31, no. 8, p. 1815, 1985.

[75] T. Damour, M. Soffel, and C. Xu, "General-relativistic celestial mechanics ii. translational equations of motion," *Physical Review D*, vol. 45, no. 4, p. 1017, 1992.

[76] A. Einstein and N. Rosen, "The particle problem in the general theory of relativity," *Physical Review*, vol. 48, no. 1, p. 73, 1935.

[77] V. A. Fok, *The theory of space, time and gravitation.* Pergamon Press Ltd, 1964.

[78] M. Friedman, *Foundations of space-time theories: Relativistic physics and philosophy of science.* Princeton University Press, 2014.

[79] N. Lexicographer, "Nasa thesaurus," NASA, Special Publication SP-2012-7501/VOL 1, 2012.

[80] T. L. Heath, *Archimedes.* Society for Promoting Christian Knowledge, 1920.

[81] V. Lenzen, "Archimedes' theory of the lever," *Isis*, vol. 17, no. 2, pp. 288–289, 1932.

[82] D. E. Smith, Archimedes, G. W. Leibniz, N. Copernicus, R. Descartes, G. Galilei, C. F. Gauss, J. Lagrange, N. Lobachevskii, J. Napier *et al.*, *Portraits of Eminent Mathematicians, with Biographical Sketches.* Scripta mathematica, Yeshiva college, 1946.

[83] L. L. Whyte, *Archimedes, or, The future of physics.* K. Paul, Trench, Trubner, 1927.

[84] R. Von Erhardt and E. von Erhardt-Siebold, "Archimedes' sand-reckoner: Aristarchos and copernicus," *Isis*, vol. 33, no. 5, pp. 578–602, 1942.

[85] A. Seiff, "Atmospheres of earth, mars, and venus, as defined by entry probe experiments," *Journal of Spacecraft and Rockets*, vol. 28, no. 3, pp. 265–275, 1991.

[86] N. Strobel, *Astronomy notes.* McGraw-Hill Learning Solutions, 2010, no. ISBN 9781121683877.

[87] R. Hu, "Atmospheric photochemistry, surface features, and potential biosignature gases of terrestrial exoplanets," Ph.D. dissertation, Massachusetts Institute of Technology, 1941.

[88] P. K. Gerard, "Planetary and satellite atmospheres," *Reports on Progress in Physics*, vol. 13, no. 1, p. 247, 1950.

[89] T. Owen, "The composition and early history of the atmosphere of mars," *Mars*, pp. 818–834, 1992.

[90] S. Seager, "Exoplanet atmospheres: Physical processes (princeton, nj," 2010.

[91] R. W. Zurek, J. R. Barnes, R. M. Haberle, J. B. Pollack, J. E. Tillman, and C. B. Leovy, "Dynamics of the atmosphere of mars," *Mars*, pp. 835–933, 1992.

[92] R. W. Zurek, "Comparative aspects of the climate of mars-an introduction to the current atmosphere," *Mars*, pp. 799–817, 1992.

[93] R. E. Sheldahl and P. C. Klimas, "Aerodynamic characteristics of seven symmetrical airfoil sections through 180-degree angle of attack for use in aerodynamic analysis of vertical axis wind turbines," Sandia National Labs., Albuquerque, NM (USA), Tech. Rep., 1981.

[94] D. Bernoulli, "Hydrodynamica," *Dulsecker. Consultable en ligne http://imgbase-scd-ulp. u-strasbg. fr/displayimage. php*, 1738.

[95] ——, "Hydrodynamica (1738)," *Translated from Latin by T. Carmody and H. Kobus, Dover pub*, p. 128, 1968.

[96] C. E. Synolakis and H. S. Badeer, "On combining the bernoulli and poiseuille equation—a plea to authors of college physics texts," *American Journal of Physics*, vol. 57, no. 11, pp. 1013–1019, 1989.

[97] D. Karwatka, "Daniel bernoulli and the airfoil lift," *Tech Directions*, vol. 69, no. 2, p. 10, 2009.

[98] M. H. Snyder Jr, "Effects of a wingtip-mounted propeller on wing lift, induced drag, and shed vortex pattern," Ph.D. dissertation, Oklahoma State University, 1967.

[99] J. Nayler, "The life of an engineer. fw lanchester: Pw kingsford. edward arnold. london. 1960. 246 pp. illustrated. 30s." *Journal of the Royal Aeronautical Society*, vol. 65, no. 605, pp. 364–364, 1961.

[100] A. Stephens, "The aeronautical work of fw lanchester," *Proceedings of the Institution of Mechanical Engineers: Automobile Division*, vol. 183, no. 1, pp. 154–166, 1968.

[101] L. Prandtl, "Theory of lifting surfaces," NACA, Technical Report 10, 1920.

[102] ——, "Applications of modern hydrodynamics to aeronautics," NACA, Technical Report 116, 1923.

[103] M. M. Munk, "The minimum induced drag of aerofoils," NACA, Technical Report TR-121, 1923.

[104] B. W. McCormick, *Aerodynamics, aeronautics, and flight mechanics*. Wiley New York, 1995, vol. 2.

[105] J. Seddon and E. L. Goldsmith, *Intake aerodynamics*. Amer Inst of Aeronautics &, 1999.

[106] J. Katz and A. Plotkin, *Low-speed aerodynamics*. Cambridge university press, 2001, vol. 13.

[107] M. O. Hansen, *Aerodynamics of wind turbines*. Routledge, 2015.

[108] P. Garabedian and D. Korn, "Numerical design of transonic airfoils," *Numerical solution of partial differential equations*, vol. 2, pp. 253–271, 1971.

[109] ——, "Analysis of transonic airfoils," *Communications on Pure and Applied Mathematics*, vol. 24, no. 6, pp. 841–851, 1971.

[110] R. T. Whitcomb, "Review of nasa supercritical airfoils," *ICAS paper*, no. 74-10, pp. 25–30, 1974.

[111] E. Omar, T. Zierten, M. Hahn, E. Szpiro, and A. Mahal, "Two-dimensional wind-tunnel tests of a nasa supercritical airfoil with various high-lift systems. volume 2: Test data," NASA, Contract Report CR2215, September 1973.

[112] H. Sobieczky and A. Seebass, "Supercritical airfoil and wing design," *Annual Review of Fluid Mechanics*, vol. 16, no. 1, pp. 337–363, 1984.

[113] B. Carmichael, "Low reynolds number airfoil survey, volume 1," NASA, Contractor Report CR-165803-VOL-1, 1981.

[114] P. Lissaman, "Low-reynolds-number airfoils," *Annual Review of Fluid Mechanics*, vol. 15, no. 1, pp. 223–239, 1983.

[115] M. Drela, "Xfoil: An analysis and design system for low reynolds number airfoils," in *Low Reynolds number aerodynamics*. Springer, 1989, pp. 1–12.

[116] P. B. MacCready, "Turbulence measurements by sailplane," *Journal of Geophysical Research*, vol. 67, no. 3, pp. 1041–1050, 1962.

[117] B. L. Pierson, "Maximum altitude sailplane winch-launch trajectories," *Aeronautical Quarterly*, vol. 28, no. 02, pp. 75–84, 1977.

[118] M. Maughmer and P. Kunz, "Sailplane winglet design," *Technical Soaring*, vol. 22, no. 4, pp. 116–123, 1998.

[119] I. Kroo, *Aerodynamics, aeroelasticity, and stability of hang gliders*. Stanford University, 1983.

[120] A. L. Thomas, "On the aerodynamics of birds' tails," *Philosophical Transactions of the Royal Society of London B: Biological Sciences*, vol. 340, no. 1294, pp. 361–380, 1993.

[121] S. A. Jenkins, D. E. Humphreys, J. Sherman, J. Osse, C. Jones, N. Leonard, J. Graver, R. Bachmayer, T. Clem, P. Carroll *et al.*, "Underwater glider system study," *Scripps Institution of Oceanography*, 2003.

[122] T. J. Mueller, *Fixed and flapping wing aerodynamics for micro air vehicle applications*. AIAA, 2001, vol. 195.

[123] R. Humble, G. Henry, and W. Larson, *Space Propulsion Analysis and Design*. McGraw-Hill, 1995.

[124] D. Shepherd, *Aerospace Propulsion*. Elsevier, 1972.

[125] G. Faeth, *Centennial of Powered Flight: A Retrospective of Aerospace Research*. AIAA, 2003.

[126] E. W. Constant, *The Origin of the Turbojet Revolution*. Baltimore: Johns Hopkins University Press, 1980.

[127] J. Golley and W. Whittle, F. ANDGunston, *Whittle: The True Story*. Shrewsbury, England: Airlife Publishing, 1987.

[128] E. Conway, *High-Speed Dreams: NASA and the Technopolitics of Supersonic Transportation, 1945–1999*. Baltimore: Johns Hopkins University Press, 2005.

[129] K. Hunekce, *Jet Engines, Fundamentals of Theory, Design and Operation*.   Osceola, WI: Motorbooks International Publishers and Wholesalers, 2003.

[130] J. Laming. (2015, March) Crosswind landings - procedures. Website. [Online]. Available: http: //code7700.com/crosswind_landing.htm

[131] I. Asimov, *Science, Numbers, and I*.   Doubleday, 1966.

[132] J. K. Beatty, C. C. Petersen, and A. Chaikin, *The new solar system*.   Cambridge University Press, 1999.

[133] U. Von Zahn, S. Kumar, H. Niemann, and R. Prinn, "13. composition of the venus atmosphere," *Venus*, p. 299, 1983.

[134] P. Gierasch, R. Goody, R. Young, D. Crisp, C. Edwards, R. Kahn, D. Rider, A. Del Genio, R. Greeley, A. Hou *et al.*, "The general circulation of the venus atmosphere: An assessment," *Venus II*, pp. 459–500, 1997.

[135] L. Zasova, N. Ignatiev, I. Khatuntsev, and V. Linkin, "Structure of the venus atmosphere," *Planetary and Space Science*, vol. 55, no. 12, pp. 1712–1728, 2007.

Index

velocity increment, 148
Venus, 58
vertical axis wind turbine, 66
vibration, 146
viscosity, 63, 75, 80
Viscous Flows, 83
volumetric radius, 165
vortex-induced lift, 66

warping, 148
water vapor, 59
wave drag, 74
webs, 145
weight distribution, 130
Weight Estimation, 19
Weight estimation, 18
Wing Loading, 105
wing sweep, 155
wings, 72
Wright Brothers , 20

X-53, 148
XB-70, 67

Yaw, 49

Zero Lift Angle of Attack, 64
zero-lift angle of attack, 66

www.ingramcontent.com/pod-product-compliance
Lightning Source LLC
Chambersburg PA
CBHW080548220326
41599CB00032B/6406